HIGH PERFORMANCE ELASTOMER MATERIALS

An Engineering Approach

AAP Research Notes on Chemistry

HIGH PERFORMANCE ELASTOMER MATERIALS

An Engineering Approach

Edited by

**Dariusz M. Bielinski, DSc, Ryszard Kozlowski, PhD,
and Gennady E. Zaikov, DSc**

Apple Academic Press

TORONTO   NEW JERSEY

Apple Academic Press Inc. | Apple Academic Press Inc.
3333 Mistwell Crescent | 9 Spinnaker Way
Oakville, ON L6L 0A2 | Waretown, NJ 08758
Canada | USA

First issued in paperback 2021

Exclusive worldwide distribution by CRC Press, a member of Taylor & Francis Group

No claim to original U.S. Government works

ISBN 13: 978-1-77463-358-8 (pbk)
ISBN 13: 978-1-77188-042-8 (hbk)

Library of Congress Control Number: 2014950114

Library and Archives Canada Cataloguing in Publication

International Science and Technology Conference ELASTOMERS "Science & Industry" (15th : 2013 : Warsaw, Poland)
High performance elastomer materials : an engineering approach / edited by Dariusz M. Bielinski, DSc, Ryszard Kozlowski, PhD, and Gennady E. Zaikov, DSc.

(AAP research notes on chemistry)
"This book presents selected papers on various aspects of rubber engineering, technology and exploitation presented during the 15th International Technological Conference ELASTOMERS 2013 "Science and Industry" held in Warsaw, Poland, 23rd–25th of October."--Preface.

Includes bibliographical references and index.
ISBN 978-1-77188-042-8 (bound)

1. Elastomers--Congresses. 2. Rubber--Congresses. I. Kozlowski, Ryszard, author, editor II. Zaikov, G. E. (Gennadiĭ Efremovich), 1935-, author, editor III. Bielinski, Dariusz M., author, editor IV. Title. V. Series: AAP research notes on chemistry

TS1925.I58 2013   620.1'94   C2014-906202-8

Apple Academic Press also publishes its books in a variety of electronic formats. Some content that appears in print may not be available in electronic format. For information about Apple Academic Press products, visit our website at **www.appleacademicpress.com** and the CRC Press website at **www.crcpress.com**

AAP RESEARCH NOTES ON CHEMISTRY

This series reports on research developments and advances in the ever-changing and evolving field of chemistry for academic institutes and industrial sectors interested in advanced research books.

Richard A. Pethrick, PhD, DSc
Research Professor and Professor Emeritus, Department of Pure and Applied Chemistry, University of Strathclyde, Glasgow, Scotland, UK

Charles Wilkie, PhD
Professor, Polymer and Organic Chemistry, Marquette University, Milwaukee, Wisconsin, USA

Georges Geuskens, PhD
Professor Emeritus, Department of Chemistry and Polymers, Universite de Libre de Brussel, Belgium

BOOKS IN THE AAP RESEARCH NOTES ON CHEMISTRY SERIES

Chemistry and Chemical Biology: Methodologies and Applications
Editors: Roman Joswik, PhD, and Andrei A. Dalinkevich, DSc
Reviewers and Advisory Board Members: A. K. Haghi, PhD, and
Gennady E. Zaikov, DSc

Functional Materials: Properties, Performance, and Evaluation
Editor: Ewa Kłodzińska, PhD
Reviewers and Advisory Board Members: A. K. Haghi, PhD,
and Gennady E. Zaikov, DSc

High Performance Elastomer Materials: An Engineering Approach
Editors: Dariusz M. Bielinski, DSc, Ryszard Kozlowski, PhD, and
Gennady E. Zaikov, DSc

ABOUT THE EDITORS

Prof. Dariusz M. Bielinski, DSc, PhD, Eng.

Prof. Dariusz M. Bielinski is affiliated with the Institute of Polymer and Dye Technology at the Faculty of Chemistry, Lodz University of Technology in Lodz, Poland, and the Division of Elastomers and Rubber Technology of the Institute for Engineering of Polymer Materials and Dyes in Piastow, Poland. He is currently involved in education and research on engineering and technology of polymer materials, especially focusing on elastomers and elastomeric composites. During his over 20-year professional career, Professor Bielinski has led a number of research projects and consulted for industry. He is the author of three monographs and 15 book chapters, over 200 scientific papers, and about 20 invited lectures on various aspects of chemistry, technology, engineering, and tribology of polymer materials and composites.

Prof. Ryszard Kozlowski, PhD, Eng.

Prof. Ryszard Kozlowski is currently affiliated with the Division of Elastomers and Rubber Technology of the Institute for Engineering of Polymer Materials and Dyes in Piastow, Poland. He is currently involved in research on natural fibers and polymer-fiber composites, especially in the aspects of their biological corrosion and flammability.

Gennady E. Zaikov, DSc

Gennady E. Zaikov, DSc, is Head of the Polymer Division at the N. M. Emanuel Institute of Biochemical Physics, Russian Academy of Sciences, Moscow, Russia, and Professor at Moscow State Academy of Fine Chemical Technology, Russia, as well as Professor at Kazan National Research Technological University, Kazan, Russia. He is also a prolific author, researcher, and lecturer. He has received several awards for his work, including the Russian Federation Scholarship for Outstanding Scientists. He has been a member of many professional organizations and is on the editorial boards of many international science journals.

CONTENTS

LIST OF CONTRIBUTORS

Arezo Afzali
University of Guilan, Rasht, Iran

Yu. O. Andriasyan
N. M. Emanuel Institute of Biochemical Physics, Russian Academy of Sciences, 4 Kosygin str., 119334, Moscow, Russia

R. Anyszka
Lodz University of Technology, Faculty of Chemistry, Institute of Polymer and Dye Technology, Stefanowskiego 12/16, 90-924 Lodz, Poland, E-mail: 800000@edu.p.lodz.pl, Tel.: +48 42 631-32-14

Dariusz M. Bieliński
Institute of Polymer and Dye Technology, Technical University of Łódź, Stefanowskiego 12/16, 90-924 Łódź, Poland, Tel.: +4842 6313214, Fax: +4842 6362543, E-mail: dbielin@p.lodz.pl; Institute for Engineering of Polymer Materials and Dyes, Division of Elastomers and Rubber Technology, Harcerska 30, 05-820 Piastow, Poland

Jacek Grams
Institute of General and Ecological Chemistry, Technical University of Łódź, Stefanowskiego 12/16, 90-924 Łódź, Poland

D. O. Gusev
Volgograd State Technical University, 400005, 28 Lenin Ave., Volgograd, Russia

A. L. Iordanskii
N. N. Semenov's Institute of Chemical Physics, RAS, Kosygin str. 4, Moscow, 119996 RF

R. Jozwik
Military Institute of Chemistry and Radiometry, Al. gen. A. Chrusciela "Montera" 105, 00-910 Warsaw, Poland, E-mail: r.jozwik@wichir.waw.pl

V. F. Kablov
Volzhsky Polytechnical Institute (branch) Volgograd State Technical University, 42a Engelsa Street, Volzhsky, Volgograd Region, 404121, Russian Federation,

P. P. Kamaev
N. N. Semenov's Institute of Chemical Physics, RAS, Kosygin str. 4, Moscow, 119996 RF

S. G. Karpova
N. M. Emanuel's Institute of Biochemical Physics, RAS, Moscow, 119996 RF

V. G. Kochetkov
Volzhsky Polytechnical Institute (branch) Volgograd State Technical University, 42a Engelsa Street, Volzhsky, Volgograd Region, 404121, Russian Federation

G. V. Kozlov
FSBEI HPE "Kh.M. Berbekov Kabardino-Balkarian State University," Chernyshevskii st., 173, Nal'chik-360004, Russian Federation

R. M. Kozlowski
Institute for Engineering of Polymer Materials and Dyes, 55 M. Sklodowskiej-Curie str., 87-100 To-run, Poland

S. V. Lapin
Volzhsky Polytechnical Institute (branch) Volgograd State Technical University, 42a Engelsa Street, Volzhsky, Volgograd Region, 404121, Russian Federation

V. S. Liphanov
Volzhsky Polytechnical Institute (branch) Volgograd State Technical University, 42a Engelsa Street, Volzhsky, Volgograd Region, 404121, Russian Federation,

N. M. Livanova
Emanuel Institute of Biochemical Physics, Russian Academy of Sciences, ul. Kosygina 4, Moscow, 119991 Russia

Shima Maghsoodlou
University of Guilan, Rasht, Iran

D. V. Medvedev
Elastomer Limited Liability Company, 400005, 75 Chuikova str. Volgograd, Russia

G. V. Medvedev
Volgograd State Technical University, 400005, 28 Lenin ave., Volgograd, Russia

I. A. Mikhaylov
N. M. Emanuel Institute of Biochemical Physics, Russian Academy of Sciences, 4 Kosygin str., 119334 Moscow, Russia

V. M. Misin
Voronezh State University of the engineering technologies N. M. Emanuel Institute of Biochemical Physics, Russian Academy of Sciences, Moscow

S. S. Nikulin
Voronezh State University of the engineering technologies N. M. Emanuel Institute of Biochemical Physics, Russian Academy of Sciences, Moscow

I. A. Novakov
Volgograd State Technical University, 400005, 28 Lenin ave., Volgograd, Russia

O. M. Novopoltseva
Volzhsky Polytechnical Institute (branch) Volgograd State Technical University, 42a Engelsa Street, Volzhsky, Volgograd Region, 404121, Russian Federation,

A. A. Olkhov
Plekhanov Russian University of Economics, Stremyanny per. 36, 117997, Moscow; Semenovl Institute of Chemical Physics, Russian Academy of Sciences, ul. Kosygina 4, Moscow, 119991 Russia

Z. Pędzich
AGH – University of Science and Technology, Faculty of Materials Science and Ceramics, Department of Ceramics and Refractory Materials, Al. Mickiewicza 30, 30-045 Krakow, Poland

P. T. Poluektov
Voronezh State University of the engineering technologies N. M. Emanuel Institute of Biochemical Physics, Russian Academy of Sciences, Moscow

A. A. Popov
Plekhanov Russian University of Economics, Stremyanny per. 36, 117997, Moscow; N.M. Emanuel Institute of Biochemical Physics, Russian Academy of Sciences, 4 Kosygin str., 119334 Moscow, Russia

K. Pyrzynski
Innovation Company, 5 Krupczyn str., 63-140 Dolsk, Poland

J. Richert
Institute for Engineering of Polymer Materials and Dyes, 55 M. Sklodowskiej-Curie str., 87-100 Torun, Poland

Mariusz Siciński
Institute of Polymer and Dye Technology, Technical University of Łódź, Stefanowskiego 12/16, 90-924 Łódź, Poland

N. V. Sidorenko
Volgograd State Technical University, 400005, 28 Lenin ave., Volgograd, Russia

Czesław Ślusarczyk
Institute of Textile Engineering and Polymer Materials, University of Bielsko-Biała, Poland

O. V. Staroverova
N. N. Semenov Institute of Chemical Physics, RAS, 119991 Moscow, street Kosygina, 4

W. Tyszkiewicz
Military Institute of Chemistry and Radiometry, 105 Allea of General A. Chrusciela, 00-910 Warsaw, Poland

M. A. Vaniev
Volgograd State Technical University, 400005, 28 Lenin ave., Volgograd, Russia, E-mail: vaniev@ vstu.ru

L. A. Vlasova
Voronezh State University of the engineering technologies N. M. Emanuel Institute of Biochemical Physics, Russian Academy of Sciences, Moscow

Z. Wertejuk
N. M. Emanuel Institute of Biochemical Physics of Russian Academy of Sciences, Moscow 119334, Kosygin st., 4, Russian Federation

Michał Wiatrowski
Department of Molecular Physics, Technical University of Łódź, Stefanowskiego 12/16, 90-924 Łódź, Poland

Kh. Sh. Yakh'yaeva
Dagestan State Pedagogical University, Makhachkala 367003, Yaragskii st., 57, Russian Federation

Yu. G. Yanovskii
Institute of Applied Mechanics of Russian Academy of Sciences, Moscow 119991, Leninskii pr., 32 a, Russian Federation

G. E. Zaikov

Military Institute of Chemistry and Radiometry, 105 Allea of General A. Chrusciela, 00-910 Warsaw, Poland; N.M. Emanuel Institute of Biochemical Physics of Russian Academy of Sciences, Kosygin str. 4, Moscow-119334, Russian Federation

Yu. N. Zernova

N. N. Semenov's Institute of Chemical Physics, RAS, Kosygin str. 4, Moscow, 119996 RF

LIST OF ABBREVIATIONS

AFM	Atomic Forced Microscopy
ALOH	Aluminium Hydroxide
BNRs	Butadiene–Acrylonitrile Rubbers
BP	British Petroleum
BSR	Butadiene-Styrene Rubber
CFM	Chloroform
C-KAO	Calcined Kaolin
CNTs	Carbon Nanotubes
CVD	Chemical Vapor Deposition
DCP	Dicumyl Peroxide
DM	Dibenzothiazole Disulphide
DSC	Differential Scanning Calorimetry
DWCNTs	Double-Walled Nanotubes
EB	Electron Beam
ENB	Ethylidene Norbornene
EP	Ethylene-Propylene Copolymer Rubber
EPDM1	Elastomer Matrix
EPDM2	Amorphous Matrix
EPDMs	Ethylene–Propylene–Diene Terpolymers
FeS	Iron Sulphide
GAB	General Aerobic Bacteria
GIXRD	Grazing Incidence X-ray Diffraction
H	Hardness
HTV	Silicone Rubber
LDPE	Low Density Polyethylene
MIC	Mica
M-MMT	Montmorillonite
NBR	Acrylonitrile-Butadiene Rubber
NR	Natural Rubber
OIT	Oxidation Induction Time
OOT	Oxidation Onset Temperature
PAH	Polycyclic Aromatic Hydrocarbons

PHA	Polyhydroxyalkanoates
PHB	Poly(3-hydroxybutyrate)
PUE	Polyurethane Elastomers
SBR	Styrene-Butadiene Rubber
SE	Secondary Electron Signal
SFE	Surface Free Energy
SIMS	Secondary Ion Mass Spectroscopy
SPIP	Scanning Probe Image Processor
SPM	Scanning Probe Microscopy
SRB	Sulfate-Reducing Bacteria
STM	Scanning Tunneling Microscopy
SWCNTs	Single-Walled Nanotubes
TC	Technical Carbon
TES	Tear Strength
TGA	Thermogravimetric Analysis
TOF	Time of Flight
TS	Tensile Strength
WOL	Wollastonite

LIST OF SYMBOLS

d_f^{if} fractal dimensions of interfacial regions

d_f^n fractal dimensions of nanofiller

$H_{nл}$ melting heat

d_f^m volume polymer matrix structure

$-\alpha$ additive values

a lower scale of polymer matrix fractal behavior

c nanoparticles concentration

$C_{¥}$ characteristic ratio

c_0 "seeds" number

d dimension of Euclidean space

D object fractal dimension

d_0 nanofiller

D_{ag} nanofiller particles aggregate diameter

D_{agr} aggregate diameter

d_f fractal dimension

D_n dimension

D_p nanofiller initial particles diameter

d_{surf} fractal dimension of nanoparticles

k Boltzmann constant

k_n proportionality coefficient

L values of long period

l_0 main chain skeletal bond length

l_{st} statistical segment length

m_0 mass of a separate particle

M_η molecular weight

N nanoparticles number per one aggregate

n_i statistical segments number in interfacial layer

Q_2 share of swelling

Q_{ad} additive value of swelling

Q_{eq}	equilibrium degree of swelling
R_{max}	nanoparticles cluster
S	cross-sectional area of macromolecule
S_{max}	ratio of maximum
S_{min}	ratio of minimum
S_n	cross-sectional area of nanoparticles
S_u	nanofiller initial particles
T	temperature
t	walk duration
T_c	temperature of crystallization
T_g	glass transition temperature
W_n	nanofiller mass contents
α	numerical coefficient
δ	solubility parameters
ΔH_c	enthalpy of crystallization
ΔH_m	enthalpy of melting
ε	exponent
ζ	increase in a number
η	medium viscosity
η_0	initial polymer
ϑ	interfacial (adsorbed) layer
ν	Poisson's ratios of composite
ν_m	Poisson's ratios of polymer matrix
ν_n	Poisson's ratios of filler
ρ_{ag}	nanofiller particles aggregate density
ρ_n	nanofiller density
τ_1	correlation times
φ_{if}	interfacial regions relative fraction
φ_n	nanofiller volume contents
χ	Flory–Huggins interaction parameter

PREFACE

This book presents selected papers on various aspects of rubber engineering, technology and exploitation presented during the 15th International Technological Conference ELASTOMERS 2013 "Science and Industry" held in Warsaw, Poland, 23rd–25th of October. A number of participants, representing the rubber industry and other industries using rubber products (such as, for example, automotive, aerospace, mining, civil engineering and transport) demonstrates the importance of elastomer materials in life today.

This book offers scope for the rubber community to present their research and development works that have potential for industrial applications as prospective materials, technologies or exploitation guidelines. The contributions range from structural and interphase phenomena in elastomer systems and new methods of the modification of filler surface and crosslinks structure of rubber vulcanizates, to modern functional elastomer composites and aspects of their thermal stability, flammability, tribology and ozone degradation. Each chapter contains a brief introduction to the particular topic, a description of the experimental techniques, and discussion of the results obtained, followed by conclusions.

Presented contributions have been divided into four parts, and namely,

Part 1: Structure of Elastomers
Part 2: Interphase Elastomer-Filler Interactions
Part 3: Modification of Elastomers and Their Components
Part 4: Stability of Elastomers

The book offers readers, both from academia and industry, the scope for broadening of their knowledge in the field of rubber compounding, crosslinking and behavior under various exploitation conditions. It should also be valuable for research students and professionals who can find theoretical background to a number of experimental techniques and their application for studying structure and related physical and chemical properties of rubber vulcanizates.

The progress in rubber engineering shown here can be the inspiration to tailoring properties of the materials during their compounding and processing as well as for prediction of exploitation performance.

PART 1
STRUCTURE OF ELASTOMERS

CHAPTER 1

SEGMENTAL MOBILITY IN CRYSTALLINE POLY(3-HYDROXYBUTYRATE) STUDIED BY EPR PROBE TECHNIQUE

A. L. IORDANSKII, P. P. KAMAEV, S. G. KARPOVA, A. A. OLKHOV, YU. N. ZERNOVA, and G. E. ZAIKOV

CONTENTS

ABSTRACT

The molecular mobility of poly(3-hydroxybutyrate) (PHB) was studied in the temperature interval from 20 to 90 °C by EPR using stable nitroxyl radicals 2,2,6,6-tetramethyl-1-piperidinyloxy (Tempo) and 4-hydroxy-2,2,6,6-tetramethyl-1- piperidinyloxy (Tempol) as spin probes. Two series of PHB samples, prepared by different methods, possessed isotropic and textured morphologies. The noncrystalline phase of PHB contain regions of two types with markedly different molecular mobilities. It is suggested that "dense" regions, characterized by a comparatively low mobility of polymer chains, are located near the surface of crystalline grains, while the "loose" (amorphous) regions with a higher mobility of chains are more distant from the surface of grains. Molecular mobility in the dense regions was virtually the same for both isotropic and textured PHB samples, whereas the mobility in the loose regions was lower in the isotropic samples than in the textured ones. Saturation of the polymer with water vapor affected both the mobility of polymer chains and the relative content of loose and dense regions in the samples.

1.1   INTRODUCTION

Poly(3-hydroxybutyrate) (PHB), the simplest and most common member of the group of polyhydroxyalkanoates (PHA) can be considered as a polymer with high potential for applications as a degradable biomedical material [1, 2]. This polymer has been extensively studied in view of its wide application in many fields as engineering, packaging, medical diagnostics, tissue engineering, drug delivery therapy and others [3–6]. Despite considerable attention of researchers toward the investigation of PHB, there are still many open questions concerning, in particular, features in molecular dynamics of this polymer in relationship with its structure [7, 8] and transport properties [9, 10]. The search for this relationship is of importance for the creation of new materials with controlled transport characteristics. The purpose of this work was to study some features of the segmental momility of PHB membranes, possessing special structural organization, by the EPR spin probe technique [11]. Another task was to determine changes in the molecular mobility of PHB upon sorption of the polymer with water vapor.

1.2   EXPERIMENTAL PART

The experiments were performed using two series of PHB samples differing by their structural organization (morphology)-textured and isotropic. The textured PHB samples were prepared by extracting the initial polymer powder (Biomer trademark) with boiling chloroform (solubility 10^{-2} g/mL). The extracted soluble fraction was used to prepare a 3% PHB solution in chloroform. The solution was poured into a Petri dish, tightly covered with glass, and allowed to stand at 20 °C until complete evaporation of the solvent. The isotropic PHB samples were prepared by dissolving the initial powder in dioxane (5%) and heating the solution to boiling. Then dioxane was completely evaporated and the residue was used to prepare a 3% PHB solution in chloroform (on heating). The solution was filtered through a Schott filter (pore size, 160), poured into a Petri dish, tightly covered with glass, and allowed to stand at 20 °C until complete evaporation of the solvent.

In order to remove the residual solvent from PHB films, the samples were kept for 2–3 h in vacuum at 80 °C. The completeness of solvent removal was checked by monitoring a decrease in intensity of the corresponding absorption bands of dioxane (873–876, and 2855 cm^{-1}) and chloroform (756, 3012–3040, 2976–2992 cm^{-1}) in the IR spectra [12]. The degree of polymer crystallinity in the samples of both types was about 70%, as evidenced by the X-ray diffraction data [7, 8]. The weight-average molecular mass determined by viscosimetry was (310 ± 26) 10^3 for the textured PHB samples and (293 ± 32) 10^3 for the isotropic ones. The nitroxyl radicals, 2,2,6,6-tetramethyl-1-piperidinyloxy (Tempo) and 4-hydroxy-2,2,6,6-tetramethyl-1-piperidinyloxy (Tempol), used as spin probes were introduced into PHB samples prior to pouring the polymer solutions into Petri dishes. The radical concentration in solid polymer samples was approx. 10^{17} spin/cm^3. The solvent-free PHB films prepared as described above were cut into 2-cm-long 2-mm-wide strips. The strip thickness varied from 60 ± 5 to 40 ± 4 μm. The EPR spectra were measured using a stack of plates placed into 4-mm-diam glass ampules. The effect of moisture on the PHB films was studied upon exposure the film samples to a saturated water vapor at 20 °C for 15–17 h. Preliminary experiments showed that this time was sufficient to provide for the equilibrium saturation of the samples with water [9]. The stacks of water-saturated samples were placed into ampules, sealed, and used to measure the EPR spectra. The water content in both isotropic and textured films under these sample preparation conditions was approx. 5.10^{-3} g/cm^3 (calculated for amorphous component), as estimated according to preliminary data on the water sorption [9, 10].

The EPR spectra were measured at the conditions far from saturation on a Radiopan SE/X-2544 spectrometer (Poland). The measurements were performed in the 20–90 °C temperature range under heating or cooling the samples at a rate of 2 K/min.

1.3   RESULTS AND DISCUSSION

The highly crystalline PHB films have formed lamellar crystals composed of macromolecules in a folded chain conformation [13]. The PHB crystals exhibit an orthorhombic unit cell with the parameters $a = 0.58$ nm, $b = 1.3$ nm, $c = 0.60$ nm [7, 8], containing two helical macromolecules with antiparallel mutual orientation [14]. In the textured PHB samples, the crystallites are oriented with their unit cell axis c along the normal to the film plane and stacked by their wide faces to form ordered domains. In the isotropic samples, the crystallites have no preferred orientation and are arranged in a random manner [7, 8]. The stable nitroxyl radicals (Tempo and Tempol) used as spin probes possess linear dimensions exceeding 0.5 nm [15–17] and cannot penetrate between the folded chains of PHB molecules forming crystallites. Thus, the radicals are most likely located in the intercrystallite layers with an average thickness of 1.8 nm [7, 8]. Indeed, the probes in crystalline polymers are usually located within disordered regions of a crystalline polymer matrix, the rotational mobility of the probes characterizing the dynamics of these regions [16, 17].

The EPR spectra of Tempo and Tempol radicals in PHB matrices (Fig. 1.1) exhibit well-resolved triplets with signs of the superposition of signals from radicals characterized by different vales of the rotation correlation time τ. The superposition is manifested by additional extrema in the main triplet signal. The spectra of probes exhibit no angular dependence. As seen from Fig. 1.1, the signs of superposition of the signals from rapidly and slowly rotating radicals are observed for the Tempol radical. Increase in the temperature is accompanied by growing intensity of the segmental motions in PHB, decreasing rotation correlation time of the spin probes, and the superposition being less pronounced as compared to the pattern observed at lower temperatures. On heating the dry PHB samples to 90 °C and their subsequent cooling, the spectra measured at equal temperatures coincide. Apparently, no irreversible structural changes take place in the polymer in the temperature range below 90 °C.

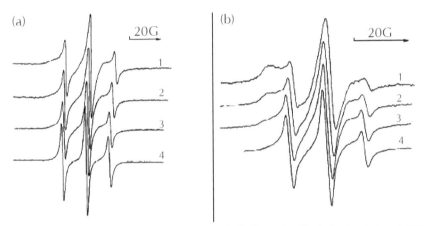

FIGURE 1.1   EPR spectra of (a) Tempo and (b) Tempol radicals in dry textured PHB films at $T = 20$ (*1*); 40 (*2*); 60 (*3*); 80 °C (*4*).

We have performed analysis of the experimental EPR spectra within the framework of the model of isotropic radical rotation. It was suggested that the observed spectra can be considered as superposition of the signals from radicals with the rotation correlation times τ_1 and τ_2, the corresponding molar fractions being w_1 and w_2. The EPR spectra of Tempo and Tempol radicals were modeled using the following spin parameters: $Azz = 34.3$ G, $Ayy = 6{,}2$ G, $Axx = 6.8$ G; $gzz = 2.00241$, $gyy = 2.00601$, $gxx = 2.00901$ [17]. The theoretical spectra were calculated using a variant of the Freed program [16] modified by Timofeev and Samariznov [18].

Figure 1.2 shows the experimental and theoretical EPR spectra of spin probes in dry and water-saturated of both isotropic and textured morphology. A comparison shows that the theoretical spectra of both Tempo and Tempol spin probes qualitatively describe the main features observed in the corresponding experimental spectra. This coincidence indicates that the spectra actually represent a superposition of the EPR signals from radical molecules with different rotation correlation times.

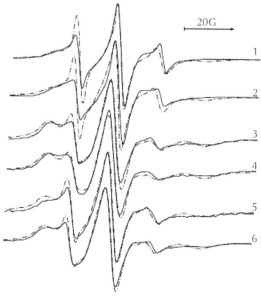

FIGURE 1.2 EPR spectra of Tempo (1, 2) and Tempol (3–6) radicals in textured (*1, 3, 5*) and isotropic (*2, 4, 6*) PHB samples in the (*1–4*) dry and water-saturated (*5, 6*) state. Dashed curves show the calculated spectra, solid lines present the experimental spectra measured at $T = 20$ °C. The curves for Tempo (*1, 2*) were calculated using the following parameters: $t_1 = 1 \times 10^{-9}$ s; $w_1 = 0.4$ (*1*), 0.25 (*2*); $t_2 = 8 \times 10^{-9}$ s (*1*), 6×10^{-9} s (*2*); $w_2 = 0.6$ (*1*), 0.75 (*2*). The parameters used in calculating the curves for Tempol (*3–6*) are listed in the table.

The existence of different rotation correlation times for the molecules of spin probes in the PHB matrix can be explained assuming that noncrystalline component of the polymer contains microscopic regions differing both in density and in the segmental mobility. One factor responsible for such a difference can be the distance from the surface of crystallites. In the vicinity of this boundary, the segmental mobility of macromolecules must be strongly hindered and the structural organization of segments can be more ordered as compared to that at a sufficiently large distance from the crystal surface. Probably, these microscopic regions are also characterized by greater density as compared to the average density of polymer in the intercrystallite layers. For brevity, these ordered regions will be referred to as the "dense" regions. On the contrary, the segments occurring at a sufficiently large distance from crystallites would possess a maximum segmental mobility and a minimum density (approaching that of the amorphous polymer). These regions will be conventionally referred to as "loose" (or amorphous). Another factor leading to a difference in behavior of the spin probe in PHB can be the presence of polar microscopic regions, formed with participation of the functional ester groups of the polymer, and less polar regions with dominating dispersion interactions. Note that the theoretical spectra of Tempol radical show a better coincidence with experiment than do the spectra calculated for the Tempo spin probe. Probably, the unsatisfactory agreement between theory and experiment for Tempo is related to the anisotropic rotation of this radical known to take place in many polymer matrices [16] but neglected in our calculation. In what follows, the consideration will be restricted to data obtained for the Tempol radical. Data on the rotation correlation time obtained by comparison of the experimental and theoretical spectra of Tempol in PHB matrices are presented in the Table 1.1 together with molar fractions of spin probes rotating with different correlation times, assigned to the microscopic regions of different types.

TABLE 1.1 Rotation Correlation Times t and Molar Fractions w of Tempol Radicals in Dense and Loose Regions of Dry and Water-Saturated PHB Samples with Different Morphologies*

Sample	$\tau_1 \times 10^9$, s	ω_1, %	$\tau_2 \times 10^9$, s	ω_2, %	$\tau_1 \times 10^9$, s	ω_1, %	$\tau_2 \times 10^9$, s	ω_2, %
	$T = 20°C$				$T = 40°C$			
Textured, dry	1.5	16	9.5	84	1.1	17	7.0	83
Textured, water-saturated	1.5	24	9.5	76	1.1	22	6.0	78
Isotropic, dry	1.9	16	9.5	84	1.5	18	7.0	82
Isotropic, water-saturated	1.3	16	8.5	84	1.4	30	6.8	70

* Subscripts 1 and 2 refer to the τ and ω values belonging to the loose (amorphous) and dense noncrystalline PHB regions, respectively.

An analysis of these data leads to the following conclusions. First, the correlation times of rotation of the Tempol spin probes located in the dense regions are virtually the same for isotropic and textured PHB samples. At the same time, the mobility of this radical in the loose regions is somewhat higher in textured matrices than in the isotropic ones. Second, the distribution of spin probes between dense and loose regions is virtually the same in both textured and isotropic PHB samples.

Saturation of the PHB samples with water vapor affects both the mobility of polymer chains (and, hence, the rotation correlation times of stable radicals localized in the microscopic noncrystalline regions of two types) and the molar ratio of the dense and loose regions. Analysis of the results obtained suggests that moisturizing of the PHB samples results in a change of the relative content of loose and dense regions in the polymer.

Figure 1.3 shows the temperature dependence of the rotation correlation time of the Tempol spin probe in PHB samples. Note that the correlation times of spin probe rotation in the dense regions coincide for the textured and isotropic samples virtually in the entire temperature range studied. At the same time, the τ values in loose regions of the textured samples are smaller as compared to those in the analogous regions of the isotropic polymer. The activation energy for the spin probe rotation in all cases was about 10 kJ/mol. The corresponding preexponential factors are 10.5×10^{-11} s for the dense regions of both textured and isotropic samples, 1.8×10^{-11} s for the loose regions of textured samples, and 3.2×10^{-11} s for these regions of isotropic samples. Our investigations [10] showed that water molecules exhibit markedly different diffusion mobility in the textured and isotropic PHB samples. The diffusion of water in the isotropic polymer is several times slower compared to that in the textured PHB of the same degree of crystallinity. The difference can be related to at least two factors. The first is a difference in the intensity of segmental motions in the intercrystallite regions of isotropic and textured polymer samples. The second is a possible significant difference in morphology (structural organization of crystallites) between the PHB samples of two types.

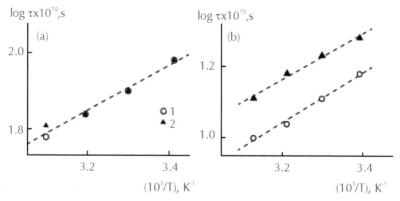

FIGURE 1.3   The plots of rotation correlation time versus temperature for Tempol radical in the (a) dense and (b) loose regions of (*1*) textured and (*2*) isotropic PHB matrices.

1.4 ACKNOWLEDGEMENTS

The work was supported by the Russian Foundation for Basic Research (grant no. 13-03-00405-a) and the Russian Academy of Sciences under the program "Construction of New Generation Macromolecular Structures" (03/OC-13).

KEYWORDS

- **Activation energy**
- **Biomedical material**
- **Isotropic polymer**
- **Macromolecular structures**
- **Molecular mobility**
- **Poly(3-hydroxybutyrate) (PHB)**

REFERENCES

1. Doi, Y. (1990). *Microbial Polyesters*, Weinheim: VCH.
2. Chodak, I. (2008). Polyhydroxyalkanoates: Origin, Properties and Applications. (In: *Monomers, Polymers and Composites from Renewable Resources*. Eds. Belgacem, M., Gandini, A.), Elsevier. NY, *566*, Ch. 22. ISBN: 978-0-08-045316-3.
3. Seebach, D., Brunner, A., Bachmann, B. M., Hoffman, T., Kuhnle, F. N. M., & Lengweiler, U. D. (1995). *Biopolymers, and Oligomers of (R)-3-hydroxyalkanoic Acids—Contributions of Synthetic Organic Chemists*. Berlin: Ernst Schering Research Foundation.
4. Artsis, M. I., Bonartsev, A. P., Iordanskii, A. L., Bonartseva, G. A., Zaikov, G. E. (2012). *Molecular Crystals and Liquid Crystals*, 555, 232–262. Biodegradation and medical application of microbial poly(3-Hydroxybutyrate).
5. Wu, Q., Wang, Y., & Chen, G.-Q. (2009). *Artificial Cells, Blood Substitutes, and Biotechnology, 37(1)*, 1–12.
6. Chen, G.-Q., & Wu, Q. (2005). *Biomaterials 26(33)*, 6565–6578.
7. Krivandin, A. V., Shatalova, O. V., & Iordanskii, A. L. (1997). *Polymer Science, Ser. B. 39(11)*, 1865.
8. Krivandin, A. V., Shatalova, O. V., & Iordanskii, A. L. (1997). *Polymer Science, Ser. B, 39(3)*, 27.
9. Iordanskii, A. L., & Kamaev, P. P. (1998). *Polymer Science, Ser. B, 40(1)*, 411.

11. Iordanskii, A. L., Krivandin, A. V., Startzev, O. V., Kamaev, P. P., & Hanggi, U. J. (1999). In: *Frontiers in Biomedical Polymer Applications*, Ottenbrite, R. M., Ed., Lancaster: Technomic Publ., *2*.
12. Kuptsov, A. H., & ZhizhinG. N. (1998). Handbook of *Fourier Transform Raman and Infrared Spectra of Polymers*, *45* (Physical Sciences Data). Elsevier. Amsterdam.
13. Seebach, D., Burger, H. M., Muller, H. M., Lengweiler, U. D., & Beck, A. K. (1994). *Helv. Chim, Acta, 77,* 1099.
14. Lambeek, G., Vorenkamp, E. J., & Schouten, A. J. (1995). *Macromolecules, 28(6)*, 2023.
15. Buchachenko, A. L., & Vasserman, A. M. (1973). *Stabil'nye radikaly* (Stable Radicals), Moscow: Khimiya.
16. Vasserman, A. M., & Kovarskii, A. L. (1986). *Spin Labels and Spin Probes in the (Physical Chemistry of Polymers)*, Moscow: Nauka, p. 139.
17. Berliner, L. J. (1976). *Spin Labeling: Theory and Applications*, Ed. New York: Academic.
18. Timofeev, V. P., & Samariznov, B. A. (1994). *Appl. Magn. Res, 4,* 523.

CHAPTER 2

INTERFACIAL LAYER IN BLENDS OF ELASTOMERS WITH DIFFERENT POLARITIES

N. M. LIVANOVA, R. JOZWIK, A. A. POPOV, and G. E. ZAIKOV

CONTENTS

ABSTRACT

On the basis of experimentally measured deviations of the equilibrium degree of swelling in n-heptane from the additive value calculated for crosslinked heterophase blends composed of butadiene–acrylonitrile rubbers of different polarities and ethylene–propylene–diene terpolymers of the known comonomer composition and stereoregularity of propylene units, the density of interfacial layer and the amount of chemical crosslinks in it have been characterized. The effects of isomers of butadiene units, the ratio of comonomers in ethylene–propylene–diene terpolymers, and the degree of isotacticity of propylene units on the intensity of interfacial interaction in covulcanizates have been analyzed.

2.1 INTRODUCTION

Formation of a strong interfacial layer is the key factor of the mechanism describing retardation of ozone degradation of a diene rubber by elastomer additives with a low degree of unsaturation [1–4]. The effect of comonomer ratio in ethylene–propylene–diene terpolymers (EPDMs) and stereoregularity of propylene units on the interfacial interaction and the amount of crosslinks in the interfacial layer was considered for heterophase crosslinked blends with butadiene–acrylonitrile rubbers (BNRs) of different polarities.

The density of the interfacial layer and the amount of crosslinks in it were determined via study of the swelling in the selective solvent n-heptane (the Zapp method [5, 6]) through deviation of the equilibrium degree of swelling from the additive value. It was proposed that the interfacial layer in the crosslinked blend of copolymers with different polarities may develop via diffusion penetration of EPDM units into the nonpolar regions of BNR [7–12].

We investigated also the density of the interfacial layer and the content of the formed crosslinks for EPDM samples with high contents of ethylene units and higher degrees of isotacticity of propylene chain fragments. It seems interesting to characterize the effect of the very low stereoregularity of propylene units, the content of diene groups, and the Mooney viscosity on the structure of interfacial region.

2.2   EXPERIMENTAL PART

The objects of research in this study were heterophase crosslinked BNR–EPDM (70: 30) blends. At this content of the nonpolar component, a system of interpenetrating crosslinked networks appears. Commercial nitrile–butadiene rubbers (trademarks BNKS-18, BNKS-28, and BNKS-40) were used. The AN-unit contents were 18, 28, and 40 wt %, respectively, and the values of the Mooney viscosity (at 100°C) were 40–50, 45–65, and 45–70 rel. units, respectively. The content of *trans*-1,4-, 1,2-, and *cis*-1,4- units of butadiene was estimated via IR spectroscopy (bands at 967, 911, and 730 cm-1) [13] with the use of extinction coefficients from [14] (Table 2.1).

TABLE 2.1   Isomeric Composition of Butadiene Units in Different Butadiene–AN Copolymers

Copolymer	Content of units, %		
	trans-1,4-	1,2-	*cis*-1,4-
BNKS -18	82.0	8.2	9.8
BNKS -28	76.4	14.4	9.2
BNKS -40	93.0	4.4	2.6

EPDM of the Royalen brand (Uniroyal, USA), of the Keltan brand the DSM 778, 714, and 712 brands (DSM N.V., Netherlands) and the domestic EPDMs having different relative amounts of ethylene, propylene, and ethylidene norbornene (ENB) units and different degrees of microtacticities of the propylene sequences, respectively, were used [15–17]. The composition, the molecular-mass characteristics, the Mooney viscosity, the isotacticity of EPDM propylene units according to IR data [13, 18, 19] are given in Tables 2.2 and 2.3.

TABLE 2.2　Composition and Basic Characteristics of Ethylene–Propylene–Diene Elastomers

EPDM Brand	Ethylene: propylene, wt %	Isotacticity, %	ENB content, wt %	$M_w \times 10^{-5}$ [9]	$M_n \times 10^{-5}$ [9]	M_w/M_n	Mooney viscosity at 125 °C
R 512	68: 32	20	4	1.95	–	1.50	57
R 505	57: 43	24	8	–	–	Narrow	55
R 521	52: 48	22	5	2.11	1.31	1.61	29
778	65: 35	13	4.5	2.0	1.35	1.48	63
714	50: 50	12	8	–	–	–	63
712	52:48	11	4,5	3.01	1.60	1.88	63

TABLE 2.3　Characteristics of Domestic EPDM

EPDM trademark	η, Mooney viscosity, rel. units	Ethylene: propylene	Isotacticity, %	ENB content, wt %
EPDM-40	36–45	70/30	29	4
Elastokam 6305	67	74/26	9.5	5.4
EPDM-60(I)	60	60/40	13	4
EPDM-60(II)	62	60/40	13	6.7
Elastokam 7505	83	60/40	9.5	5.1

For domestic EPDMs the data on the Mooney viscosity; the content of ethylene, propylene, and ethylidenenorbornene units was according to the manufacturer data.

A vulcanizing system for NBRs had the following composition, phr: stearic acid, 2.0; Sulfenamide Ts (N-cyclohexylbenzothiazole-2-sulfenamide), 1.5; zinc oxide, 5.0; and sulfur, 0.75. EPDM of the Royalen brand, the DSM and domestic EPDM was vulcanized with supported Peroximon F-40 taken in an amount of 5.5 phr. Each rubber was mixed with its vulcanizing system by roll milling at 40–60 °C for 15 min. Then, a rubber blend was prepared under the same conditions. The blends were vulcanized at 170 °C within 15 min.

The density of the interfacial layer and the amount of crosslinks in this layer were characterized by calculation of the deviation of the equilibrium degree of swelling Q_{eq} from additive values toward increase in the nonpolar solvent n-heptane [5, 6]. The deviation is related to a weak interfacial interaction between thermodynamically incompatible polymer components, one of which contains polar units. In such systems, as was suppose, only local segmental solubility of nonpolar chain portions is possible [7–12].

The Flory–Huggins interaction parameter χ for polybutadienes and EPDM with n-heptane and solubility parameters δ or cis-PB, EPDM, and BNKS were reported in [20, 21].

The deviation of Q_{eq} from additive values $-\alpha$ was calculated by the formula [6]:

$$-\alpha = [(Q_{ad} - Q_{eq})/(Q_{ad} - Q_2)] \times 100\%,$$

where Q_{eq} is the equilibrium degree of swelling of a covulcanizate; Q_{ad} is the additive value of swelling in a given solvent, as calculated from the equilibrium degree of swelling of vulcanizates for each rubber; and Q_2 is the share of swelling of the second elastomer (BNR).

2.3   RESULTS AND DISCUSSION

Calculations showed that the value of $-\alpha$ for BNRS-28 is somewhat lower than that for BNRS-18, even though the polarity of the former rubber is higher. This phenomenon was attributed to the effect of various butadiene isomers on compatibility with ethylene–propylene–diene elastomers (Table 2.1).

The value of $-\alpha$ depends on the amount of crosslinks in the interfacial layer and its volume. Let us assume that the major fraction of polar units of BNR is uninvolved in its formation and the value of $-\alpha$ was recalculated to the 100% content of butadiene units $-\alpha^{100\%}$ (Fig. 2.1) and used to characterize the structure of the interfacial interaction zone and the amount of crosslinks contained in it. In such a manner, the effect of the interfacial layer volume could be minimized. As will be shown below, this situation may not be attained in all cases.

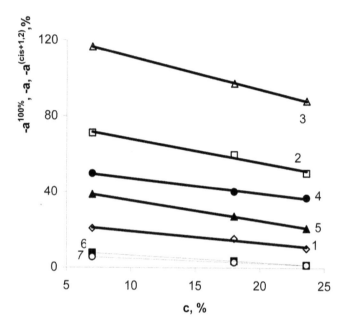

FIGURE 2.1 The plots the value of $-\alpha 100\%$ for covulcanizates EPDM with DSM (*1*) 778, (*2*) 714, (*3*) 712, (*4*) EPDM-60(I); $-\alpha$ (*5*) for R 521; $-\alpha^{cis+1,2}$ for (*6*) R 505 and (*7*) Elastokam 6305 as a function of the total content of 1,4-*cis*- and 1,2-butadiene isomers in BNR.

It is seen that a linear decrease in the value of $-\alpha^{100\%}$ with an increase in the total content of 1,4-*cis* and 1,2 units is observed for the crosslinked blends of BNR with all DSM EPDM samples and EPDM-60(I) characterized by a low isotacticity of propylene (Tables 2.2 and 2.3). The fact that the value of $-\alpha^{100\%}$ decreases in proportion to the amount of 1,4-*cis* and 1,2 isomers of butadiene units for EPDM-based blends provides evidence that the density of the transition layer increases. This circumstance implies that the mutual solubility of EPDM comonomers and butadiene units that occur for the most part in the 1,4-*trans* configuration in the neighborhood of these isomers is improved.

For DSM and EPDM-60(I) the region of interfacial interaction is bounded by nonpolar BNR units. The compatibility of chain portions of these EPDM samples with the polar acrylonitrile groups is ruled out. As the proportion of propylene units in EPDM is increased, the compatibility of the components, the density of the interfacial layer, and the amount of crosslinks in it drop sharply (the absolute value of $-\alpha^{100\%}$ increases). Thus, the higher the content of atactic propylene units in EPDM has the lower the adhesion interaction of the components and the worse its compatibility with BNR. The strengthening of interfacial interaction with an increase in Σ(1,4-*cis* and 1.2 units) for EPDM 778, EPDM 714, and EPDM 712 by 2.0, 1.4, and 1.3 times, respectively, is apparently explained by the loosening effect of these isomers on the structure of the nonpolar phase of BNR comprising for the most part 1,4-*trans* units (Table 2.3), which show the tendency toward ordering at the low content of acrylonitrile and other isomers [22–24]. Ordered structures worsen compatibility of polymers even to a higher extent [7–12, 24].

When the Uniroyal EPDM, which is characterized by a high stereoregularity of propylene sequences, was tested in blends, no proportionality between the values under consideration was observed (Fig. 2.2) [25].

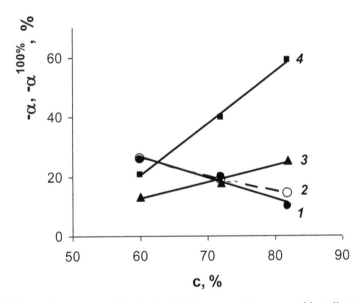

FIGURE 2.2   *(1,4)* *(–α)* and *(2, 3)* *(–α100%)* vs. overall content of butadiene units *c* for *(1)* EPDM R 512, *(2)* EPDM-40, *(3)* EPDM-60 (II), *(4)* Elastokam 7505, in their covulcanizates with BNR.

For BNR covulcanizates with EPDM R 512 containing a large proportion of ethylene units and distinguished by the presence of stereoregular propylene sequences, the value of $-\alpha^{100\%}$ linearly decreases with the content of butadiene units by a factor of 2.6 (Fig. 2.2). This fact leads us to infer that, firstly, ethylene units adjoining predominantly short isotactic propylene sequences are well compatible with all isomers of butadiene units. Secondly, an increase in the density of the interfacial layer and in the amount of crosslinks in it (a decrease in $-\alpha^{100\%}$) with a rise in the proportion of non-polar units in BNR implies that EPDM molecular fragments may penetrate into BNR regions that apparently contain single polar acrylonitrile groups.

Figure 2.2 presents the $(-\alpha)$ value plotted against the content of butadiene units in BNR for the crosslinked blends with EPDM-40 (curve 2). As compared with other commercial EPDMs (Tables 2.2 and 2.3), EPDM-40 is characterized by a very high degree of isotacticity (29%) of propylene chain fragments (about 90% of the isotactic fraction). As was shown propylene fragments with a high stereoregularity are able to penetrate into the BNR regions that are likely to contain single polar groups so as for EPDM R 512.

EPDM-40 with a higher degree of isotacticity of propylene units (as compared with EPDM R 512) is characterized by a far more intensive penetration of propylene chain fragments in their rigid isotactic configuration into BNR regions containing single polar groups. An increased volume of the interfacial layer and its local density reduction are confirmed by the fact that the value of $(-\alpha)$ for blends with EPDM-40 is higher than that for the EPDM R 512-based blends. This increase comes to 1.3-, 1.5-, and 1.7-folds for BNKS-18, BNKS-28, and BNKS-40, respectively.

For the covulcanizates of R 521 EDPM, which contains a large amount of isotactic propylene sequences [15, 16, 25], the proportional decrease in the value of $-\alpha$ by a factor of 1.85 is observed in the $-\alpha - \Sigma(1,4\text{-}cis$ and 1.2 units) coordinates (Fig. 2.1). Consequently, at a high content of isotactic propylene units, EPDM shows better compatibility with BNR if the butadiene comonomer is enriched with 1,4-cis and 1.2 units (a reduction in the value of $-\alpha$). The values of $-\alpha$ do not take into consideration restrictions related to the mutual interpenetration of segments of dissimilar chains associated with the presence of polar groups in BNR. As a result, it is not improbable that BNR portions containing polar groups may be involved in the interfacial interaction region. This circumstance is related to the specific features of BNR interaction with EPDM containing a large proportion of isotactic propylene chain fragments. This observation may be attributed to the rigidity of isotactic propylene sequences arising from hindered conformational transitions. The potential barrier to transitions between rotational isomers of monomer units for the isotactic PP is 21 kJ/mol, while for PE the potential barriers of $T–G$ and $G–G$ transitions are 2.5 and 8.8−10 kJ/mol, respectively [26].

For blends with EPDM R 505, which contains a large amount of ENB, the linear dependence is attained when the values of $-\alpha^{100\%}$ are recalculated to the 1% of the sum of 1,4-cis and 1.2 units in BNR ($-\alpha^{cis+1.2}$) (Fig. 2.1) [2]. An analysis of the $-\alpha^{cis+1.2}$ versus Σ(1,4-cis and 1.2 units) curves demonstrates that $-\alpha^{cis+1.2}$ decreases by a factor of 5.3 with an increase in the total amount of these butadiene unit isomers. The data presented above suggest that the chain fragments of this EPDM are well compatible only with those portions of butadiene chains that contain 1,4-cis and 1,2 isomers, while in the case of 1,4-$trans$ units, compatibility is much worse. It appears that the bulky diene group, as in EPDM 714, significantly hinders the incorporation of EPDM chain portions into butadiene regions of BNR.

For Elastokam 6305 with a high content of ethylene units (74%), with a higher amount of ENB (5.4%) and a very low degree of isotacticity of propylene sequences (9.5%) (Table 2.2), as follows from Fig. 2.1, the dependence on the overall content of cis-1,4- and 1,2-butadiene units in BNR is linear if the $(-\alpha)$ value is normalized to the 100% content of nonpolar units and 1% content of cis-1,4- and 1,2-isomers ($-\alpha^{cis-1.4+1.2}$). Therefore, one can conclude that this copolymer is compatible only with butadiene chain fragments of BNR containing preferably cis-1,4- and 1,2-isomers. Hence, when propylene units are characterized by a very low microtacticity (9.5%), compatibility between the components decreases. As was shown in [25], the high content of ethylene units, whatever the configuration of propylene sequences, provides a better compatibility with butadiene copolymer BNKS. However, as atactic configuration in the blend dominates, the depth of penetration of EPDM segments is limited by the butadiene part of the copolymer.

Because, in EPDM-60 (I), half of all propylene units exist in the isotactic configuration and the content of diene groups is lower, its segments are able to diffuse into nonpolar SKN regions to a greater depth than segments of Elastokam 6305.

EPDM-60 (II) with a higher content of ENB (6.7%) is compatibilized only with butadiene units (Fig. 2.2, curve 2) and $(-\alpha^{100\%})$ increases by a factor of 1.8 as their content grows. This tendency suggests that this interfacial layer is characterized by an increased friability and the number of crosslinks in this layer is low. The higher the volume of this layer (the lower the polarity of BNR) the higher the $-\alpha^{100\%}$ values.

The ratio between ethylene and propylene units in Elastokam 7505 is similar to that in EPDM-60 (II), but the degree of isotacticity of propylene units is lower (9.5%) and the Mooney viscosity is very high (Table 2.3). Elastokam 7505 is characterized by the minimum compatibility with all BNR samples. As follows from Fig. 2.2 (curve *4*), the friability of the interfacial layer of this EPDM is maximum; as a result, $(-\alpha^{100\%})$ markedly increases with a decrease in the polarity of SKN (by a factor of ~3). The worst results were observed for EPDM-60 (II) with an increased content of diene units and for Elastokam 7505 with a high Mooney viscosity.

Thus, the structure of interfacial interaction zone depends on the comonomer composition of EPDM, the stereoregularity of propylene units, and the isomerism of butadiene units.

Owing to conformational restrictions, stereoregular propylene units can penetrate into BNR regions that apparently contain single polar acrylonitrile units. As a result, the volume of the interfacial zone increases but its density and the amount of crosslinks contained in it decrease locally. The compatibility of propylene fragments of EPDM chains that predominantly occur in the atactic configuration with butadiene units is much worse. The interfacial layer is looser, and the region of diffusion penetration of phases is confined by the presence of nonpolar units. The proportional growth of the density of interfacial layer and the amount of crosslinks with an increase in the total amount of isomers of butadiene units contained in BNR in a smaller amount is explained by disordering of 1,4-*trans* units. As a result, their compatibility with portions of EPDM molecules is facilitated.

The ability of EPDM that contains a large amount of diene units and stereoregular propylene units to compatibilizer with butadiene units of BNR is close to that of EPDM with a low degree of isotacticity of the propylene comonomer and is determined by steric hindrances related to the presence of the bulky diene. However, a higher content of diene groups ensures better crosslinking of EPDM with the matrix. As a result, the total amount of crosslinks between phases is higher than that for blends with EPDM having the same ratio of ethylene and propylene units but a smaller amount of diene (cf. EPDM R 505 and R 521, EPDM 714 and EPDM 712).

The largest amount of crosslinks in the interfacial layer forms when EPDM with a high content of ethylene units is used despite the moderate amount of diene contained in it.

KEYWORDS

- **Butadiene–nitrile rubbers**
- **Compatibility**
- **Ethylene–propylene–diene elastomers**
- **Interphase layers**
- **Isomeric composition**
- **Isotacticity propylene**
- **Phase structure**
- **Stereoregularity**
- **Units**

REFERENCES

1. Krisyuk, B. E., Popov, A. A., Livanova, N. M., & Farmakovskaya, M. P. (1999). *Polymer Science, Ser.* A 41, 94 (1999). [*Vysokomol. Soedin., Ser. A, 41,* 102].
2. Livanova, N. M., Lyakin, Yu. I., Popov, A. A., & Shershnev, V. A. (2007). *Polymer Science, Ser.* A, 49, 63, [*Vysokomol. Soedin., Ser.* A 2007, *49,* 79].
3. Livanova, N. M., Lyakin, Yu I., Popov, A. A., & Shershnev, V. A. (2007). *Polymer Science, Ser.* A 49, 300 (2007). [*Vysokomol. Soedin., Ser.* A, *49,* 465].
4. Livanova, N. M., Lyakin, Yu I., Popov, A. A., & Shershnev, V. A. (2006). *Kauch. Rezina, 4,* 2.
5. Gould, R. F. (1972). *Multicomponent Polymer Systems,* Ed. by American Chemical Society, Washington.
6. Lednev, Yu. N., Zakharov, N. D., & Zakharkin, O. A., et al. (1977). *Kolloidn. Zh, 39,* 170.
7. Kuleznev, V. N. (1980). *Polymer Blends* (Khimiya, Moscow) [in Russian].
8. Lipatov, Yu. S. (1986). *Physical Chemistry of Multicomponent Polymer Systems,* Ed. by (Naukova Dumka, Kiev). 2 [in Russian].
9. Kuleznev, V. N. (1993). *Vysokomol. Soedin., Ser.* B, 35, 1391 [*Polymer Science, Ser. B* 1993, 35, 839].
10. Kuleznev, V. N., & Voyutskii, S. S. (1973). *Kolloidn. Zh, 35,* 40.
11. Orekhov, S. V., Zakharov, N. D., Kuleznev, V. N., & Dogadkin, B. A. (1970). *Kolloidn. Zh., 32,* 245.

12. Grishin, B. S., Tutorskii, I. A., & Yurovskaya, I. S. (1978). *Vysokomol. Soedin., Ser.* A, *20*, 1967.

13. Dechant, J., Danz, R., Kimmer, W., & Schmolke, R. (1972). *Ultrarotspektroskopische Untersuchungen an Polymeren* (Akademie, Berlin; Khimiya, Moscow, 1976).

14. Kozlova, N. V., Sukhov, F. F., & Bazov, V. P. (1965). *Zavod. Lab. 31*, 968

15. Livanova, N. M., Karpova, S. G., & Popov, A. A. (2003). *Polymer Science, Ser.* A, 53, 1128 [*Vysokomol. Soedin., Ser.* A 2003, 53, 2043].

16. Livanova, N. M., Evreinov, Yu. V., Popov, A. A., & Shershnev, V. A. (2003). *Polymer Science, Ser.* A 45, 530 [*Vysokomol. Soedin., Ser.* A 2003, *45*, 903].

17. Livanova, N. M., Karpova, S. G., & Popov, A. A. (2005). *Plast. Massy*, *2*, 11.

18. Kissin, Yu. V., Tsvetkova, V. I., & Chirkov, N. M. (1963). *Doklady AN SSS, 152*, 1162.

19. Kissin, Yu. V., Popov, I. T., & Lisitsin, D. M., et al. (1966). *Proizvod. Shin Rezino-Tekh. Asbesto-Tekh. Izdeli*, *7*, 22.

20. Saltman, W. M. (1979). *The Stereo Rubbers*, (Wiley, New York), 2.

21. Nesterov, A. E. (1984). *Handbook on Physical Chemistry of Polymers* (Naukova Dumka, Kiev) [in Russian].

22. Bartenev, G. M. (1979). *Structure and Relaxation Properties of Elastomers* (Khimiya, Moscow) [in Russian].

23. Livanova, N. M., Karpova, S. G., & Popov, A. A. (2009). *Polymer Science, Ser.* A 2009, 51, 979 [*Vysokomol. Soedin., Ser.* A, *51*, 1602].

24. Livanova, N. M., Karpova, S. G., & Popov, A. A. (2011). *Polymer Science, Ser.* A 2011, 53, 1128 [*Vysokomol. Soedin., Ser.* A, *53*, 2043].

25. Livanova, N. M. (2006). *Polymer Science, Ser.* A, *48*, 821 [*Vysokomol. Soedin., Ser.* A 2006, *48*, 1424].

26. Vol'kenshtein, M. V. (1959). *Configurational Statistics of Polymer Chains* (Akad. Nauk SSSR, Moscow) [in Russian].

CHAPTER 3

INFLUENCE OF MOLECULAR STRUCTURE OF COMPONENTS ON CRYSTALLINITY AND MECHANICAL PROPERTIES OF LDPE/EPDM BLENDS

DARIUSZ M. BIELIŃSKI, and CZESŁAW ŚLUSARCZYK

CONTENTS

ABSTRACT

Structural aspects of components constituting low density polyethylene/ethylene-propylene-diene rubber (LDPE/EPDM) blends are studied in bulk and compared to the surface layer of materials. Solvatation of a crystalline phase of LDPE by EPDM takes place. The effect is more significant for systems of amorphous matrix, despite a considerable part of crystalline phase in systems of sequenced EPDM matrix seems to be of less perfect organization. Structural data correlate perfectly with mechanical properties of the blends. Addition of LDPE to EPDM strengthens the material. The effect is higher for sequenced EPDM blended with LDPE of linear structure.

The surface layer exhibits generally lower degree of crystallinity in comparison to the bulk of LDPE/EPDM blends. The only exemption is the system composed of LDPE of linear structure and amorphous EPDM exhibiting comparable values of crystallinity. Microindentation data present the negative surface gradient of hardness. Sequenced elastomer matrix always produces significantly lower degree of crystallinity, no matter LDPE structure, whereas systems of amorphous EPDM matrix follow the same trend only when branched polyethylene of lower crystallinity is added. Values of long period (L) for the blends are significantly higher than that of their components, what suggest some part of ethylene sequences from elastomer phase to take part in recrystallization. LDPE of linear structure facilitates the phenomenon, especially if takes in amorphous EPDM matrix. Branched LDPE recrystallizes to the same lamellar thickness, no matter the structure of elastomer matrix.

3.1 INTRODUCTION

Polyolefine blends are group of versatile materials, which properties can be tailored to specific applications already at the stage of compounding and further processing. Our previous papers on elastomer/plastomer blends were devoted to phenomenon of cocrystallization in isotactic polypropylene/ethylene-propylene-diene rubber (iPP/EPDM) [1] or surface segregation in low density polyethylene/ethylene-propylene-diene rubber (LDPE/EPDM) [2, 3] systems. Composition and structure of the materials were related to their properties. Recently, we have described the influence of molecular weight and structure of components on surface segregation of LDPE in blends with EPDM, and morphology of the surface layer being formed [4].

This paper completes the last one with structural data, calculated from X-rays diffraction spectra, collected for bulk and for the surface layer of LDPE/EPDM blends. We have focused on comparison between mechanical properties, relating them to the degree of crystallinity and a crystalline phase being formed in bulk and in the surface layer of the systems.

3.2 EXPERIMENTAL PART

3.2.1 MATERIALS

The polymers used in this study are listed in Table 3.1, together with their physical characteristics.

TABLE 3.1   Physical Characteristics of the Polymers Studied

Polymer	Density [g/cm^3]	Solubility parameter [J$^{0.5}$/m$^{1.5}$][1]	Degree of branching FTIR[2]	Melting temperature [°C]	Molecular weight, M_w	Dispersity index M_w/M_n
EPDM1	0.86	15.9×10^{-3}	-	44	-	-
EPDM2	0.86		-	-	-	-
LDPE1	0.930	15.4×10^{-3}	3.8	112	15,000	2.32
LDPE2	0.906		6.0	90	35,000	2.56

EPDM1, EPDM2 – ethylene-propylene-diene rubber: Buna EPG-6470, G-3440 (Bayer Germany; monomer composition by weight: 71% and 48% of ethylene, 17% and 40% of propylene, 1.2% of ethylideno-norbornene, respectively).
LDPE1, LDPE2 – low density polyethylenes (Aldrich Chemicals, UK: cat. no. 42,778–0 and 42,779–9 respectively).
[a] Taken from Ref. [5].
[b] Determined from infrared spectra according to Ref. [6].

Low density polyethylene (LDPE) in the amount of 15 phr was blended with ethylene-propylene-diene terpolymer (EPDM). The method of blend preparation, at the temperature of 145 °C, that is, well above melting point of the crystalline phase of polyethylene, was described earlier [2]. To cross-link elastomer matrix 0.6 phr of dicumyl peroxide (DCP) was admixed to the system during the second stage at 40 °C. The systems were designed in a way enabling studying the influence of molecular structure of the rubber matrix or molecular weight, crystallinity/branching of the plastomer on mechanical properties of LDPE/EPDM blends, both: in bulk and exhibited by the surface layer. Structural branching of the polyethylenes studied was simulated from their ^{13}C NMR spectra applying the Cherwell Scientific (UK) NMR software. LDPE1 contains short branches, statistically every 80 carbons in the backbone. It was found that every fourth branch is longer, being constituted of 6–8 carbon atoms. LDPE2, characterized by higher degree of branching, contains however short branches of 2–4 carbon atoms, but placed ca. every 15 backbone carbons. Samples were steel mold vulcanized in an electrically heated press at 160 °C, during time $t_{0.9}$ determined rheometrically, according to ISO 3417.

3.2.2 TECHNIQUES

3.2.2.1 X-RAYS DIFFRACTION

WAXS and SAXS measurements were carried out applying the same equipment and procedures described previously [1].

3.2.2.1.1 WAXS

Investigations were performed in the scattering angle range 5–40° with a step of 0.1°. Each diffraction curve was corrected for polarization, Lorentz factor and incoherent scattering. Each measured profile was deconvoluted into individual crystalline peaks and an amorphous halo, following the procedure described by Hindeleh and Johnson [7]. Fitting was realized using the method proposed by Rosenbrock and Storey [8]. The degree of crystallinity was calculated as the ratio of the total area under the resolved crystalline peaks to the total area under the unresolved X-rays scattering curve.

3.2.2.1.2. SAXS

Measurements were carried out over the scattering angle $2\Theta=0.09-4.05°$ with a step of 0.01° or 0.02°, in the range up to and above 1.05°, respectively. Experimental data were smoothed and corrected for scattering and sample absorption by means of the computer software FF SAXS-5, elaborated by Vonk [9]. After background subtraction, scattering curves were corrected for collimation distortions, according to the procedure proposed by Hendricks and Schmidt [10, 11]. Values of the long period (L), were calculated from experimental data, using the Bragg equation:

$$L=\lambda/(2 \sin\Theta) \qquad (1)$$

where: L – long period, Θ – half of the scattering angle, λ – wavelength of the X-rays.

3.2.2.1.3 GIXRD

Grazing incidence X-ray diffraction (GIXRD) experiments were carried out at room temperature using a Seifert URD-6 diffractometer equipped with a DSA 6 attachment. CuKα radiation was used at 30 kV and 10 mA. Monochromatization of the beam was obtained by means of a nickel filter and a pulse-height analyzer. The angle of incidence was fixed to 0.5°, so that the X-rays penetration into the sample could be kept constant during measurements. Diffraction scans were collected for 2θ values from 2° to 60° with a step of 0.1°.

3.2.2.2 DIFFERENTIAL SCANNING CALORIMETRY (DSC)

Enthalpies of melting were determined with a NETZSCH 204 differential scanning calorimeter (Germany), calibrated for temperature and enthalpy using an indium standard. Specimens of about 9–10 mg were frame cut from sheets of a constant thickness to eliminate possible influence of the specimen geometry on a shape of DSC peak. Experiments were carried out over the temperature range 30–160 °C. Prior to cooling down, the samples were kept for 5 min at 160 °C. Melting and crystallization were carried out with a scanning rate of 10°/min. Temperature of melting – Tm or crystallization – T_c were taken as the ones corresponding to 50% of the adequate transition. The enthalpy of melting – ΔH_m or crystallization – ΔH_c were taken as areas under the melting or crystallization peak, respectively. The degree of the blend crystallinity was calculated from a ΔH_m value, according to the formula:

$$X_C = \frac{\Delta H_m}{\Delta H_m^\circ} \qquad (2)$$

where ΔH°_m stands for the enthalpy of melting of polyethylene crystal: 289 J/g [12].

3.2.2.3   MECHANICAL PROPERTIES

Mechanical properties of polymer blends during elongation were determined with a Zwick 1435 instrument, according to ISO 37. Dumb-bell specimen geometry number 2 was applied.

3.2.2.4   NANOINDENTATION

Hardness and mechanical moduli of polymer blends were determined with a Micromaterials Nano Test 600 apparatus (UK), applying the procedure of spherical indentation with 10% partial unloading. R=5 μm stainless steel spherical indenter probed the surface layer of material with the loading speed of dP/dt=0.2 mN/s, reaching depths up to 8.0 μm. More information on the instrumentation can be found elsewhere [13].

3.3   RESULTS AND DISCUSSION

3.3.1   SUPERMOLECULAR STRUCTURE IN BULK

Degree of bulk crystallinity, calculated for the blends from DSC and WAXS spectra, are given in Table 3.2.

TABLE 3.2   Degree of Bulk Crystallinity of the Materials Studied

Sample	Degree of crystallinity [%]			
	WAXS exp.	WAXS calc.	DSC exp.	DSC calc.
EPDM1	20.5	–	3.9	–
EPDM2	0	–	0	–
LDPE1	47.3	–	61.4	–
LDPE2	30.6	–	32.0	–
LDPE1/EPDM1	22.4	24.0	8.4	11.4
LDPE2/EPDM1	19.0	21.8	5.0	10.0
LDPE1/EPDM2	2.7	6.2	4.8	8.0
LDPE2/EPDM2	1.6	4.0	2.8	4.2

They differ significantly from the values calculated additively. Solvatation of a crystalline phase of plastomer by elastomer matrix seems to be apparent. The effect, exceeding 50%, is significant in the case of amorphous matrix (EPDM2), whereas in the case of sequenced elastomer matrix (EPDM1) the difference between degree of crystallinity calculated additively and determined by WAXS is less than 10%, approaching experimental error.

DSC spectra of LDPE/EPDM blends follow the WAXS analysis in the case of system with amorphous matrix (EPDM2). For the blends of sequenced elastomer matrix (EPDM1) the degree of crystallinity calculated from heat of melting data is significantly lower than determined from X-rays wide angle diffraction and points definitely on the crystalline phase swelling phenomenon to take place. In our opinion the huge difference in the degree of crystallinity of EPDM1 between calculations from DSC (X_c=3.9%) and WAXS (X_c=20.5%) spectra is responsible for the former, whereas limited resolution of calorimetry to less perfect organization of macromolecules (paracrystalline phase) seems to be a justification for the latter. Packing of macromolecules, is of high order for the plastomers studied, whereas for sequenced elastomer seems to be quite low, judging from the mentioned already huge difference on X_c of EPDM1, depending on the applied method of analysis.

3.3.2 MECHANICAL PROPERTIES

Values of stress at 100% (SE_{100}), 200% (SE_{200}), 300% (SE_{300}) of elongation, tensile strength (TS) and elongation at break (EB) of the composites are very similar, except M-MMT sample filled with surface modified
montmorillonite (Table 3.3). In its case values of moduli at 100, 200 and
300% of elongation and tensile strength were slightly lower than for other
composites, but its elongation at break was almost two times higher than
determined for the other samples. Also M-MMT sample characterize itself
by the best tear strength higher from 19% up to even 90% than composites
filled with unmodified refractory powders. The highest value of hardness
has the composite filled with wollastonite (WOL) whereas the lowest the
composite containing surface modified montmorillonite (M-MMT). However the difference is not significant – it does not exceed ca 10%.

Mechanical properties of the blends are given in Table 3.4.

TABLE 3.3   Mechanical Properties of the Composites Studied

Composites Properties	C-KAO	WOL	ALOH	M-MMT	MIC
SE_{100} [MPa]	1.9	1.7	1.7	1.3	2.0
SE_{200} [MPa]	2.8	2.6	2.6	1.5	2.7
SE_{300} [MPa]	3.7	3.5	3.4	1.8	3.4
TS [MPa]	3.9	3.9	3.7	3.1	3.9
EB [%]	322	350	338	641	357
TES [N/mm]	8.2	10.4	9.3	15.6	13.1
H [°ShA]	63.1	68.0	66.2	62.4	64.1

Values of Stress at 100% (SE_{100}), 200% (SE_{200}) and 300% (SE_{300}) of Composites Elongation
and their Tensile Strength (TS), Tear Strength (TES), Elongation at Break (EB) and
Hardness (H).

TABLE 3.4   Mechanical Properties of the Materials Studied

Sample	S_{100} [MPa]	S_{200} [MPa]	S_{300} [MPa]	TS [MPa]	E_b [%]
EPDM1	3.3	4.5	6.0	6.9	508
EPDM2	1.0	1.2	1.4	4.0	666
LDPE1/EPDM1	3.7	4.1	4.7	20.4	555
LDPE2/EPDM1	3.1	3.6	4.5	17.5	902
LDPE1/EPDM2	1.7	1.9	2.0	4.9	852
LDPE2/EPDM2	1.6	1.8	1.9	4.3	792

They correlate well with structural data. The higher the degree of crystallinity the higher the moduli in extension, tensile strength of material and elongation at break. Addition of LDPE to EPDM improves mechanical properties of the elastomer, what is especially visible for the sequenced matrix. Strengthening effect is more pronounced when lower molecular weight but of higher crystallinity plastomer is added.

3.3.3   SUPERMOLECULAR STRUCTURE IN THE SURFACE LAYER

Degree of crystallinity of the surface layer of blends, calculated from WAXS spectra (low incidence angle) are given in Table 3.5.

TABLE 3.5   Degree of the Surface Layer Crystallinity of the Materials Studied

Sample	Degree of crystallinity [%]	
	WAXS exp.	WAXS calc.
EPDM1	9.6	–
EPDM2	0	–
LDPE1	47.3	–
LDPE2	32.0	–
LDPE1/EPDM1	10.5	14.5
LDPE2/EPDM1	6.1	12.3
LDPE1/EPDM2	5.6	6.2
LDPE2/EPDM2	1.3	4.0

The surface layer of both polymer components and their blends, exhibits generally, in comparison to the bulk, lower values of the degree of crystallinity. The only exemption is the LDPE1/EPDM2 system, for which the crystallinity of the surface layer is significantly higher than in bulk and additive calculations give the value only slightly higher in comparison to the experimental one.

Comparing the WAXS data determined in bulk to the ones characterizing the surface layer of the systems studied, one can find that their relation does depend on supermolecular structure of components. Sequenced elastomer matrix always produces significantly lower than in bulk degree of crystallinity, no matter the structure of plastomer, whereas the same is followed by amorphous elastomer matrix only when branched polyethylene (LDPE2) of lower crystallinity is added. Amorphous EPDM matrix facilitates crystallization of low molecular weight polyethylene of higher crystallinity (LDPE1) on the surface.

Values of long period (L) for the blends are significantly higher than that of their components (Table 3.6).

TABLE 3.6　Values of the Long Period (L) of the Materials Studied (SAXS)

Sample	L [nm]
EPDM1	14.3
EPDM2	0
LDPE1	14.5
LDPE2	10.6
LDPE1/EPDM1	17.8
LDPE2/EPDM1	14.2
LDPE1/EPDM2	24.4
LDPE2/EPDM2	14.2

It suggest some part of ethylene sequences from the elastomer phase to take part in recrystallization. Polyethylene of higher linearity (LDPE1) facilitates the phenomenon, which takes place to the higher extent in the amorphous elastomer matrix (EPDM2). Plastomer of higher branching (LDPE2), recrystallizes to the same lamellar thickness, no matter the structure of the elastomer matrix.

3.3.4   NANOINDENTATION

Hardness profiles of the surface layer of blends studied are presented in Figs. 3.1 and 3.2.

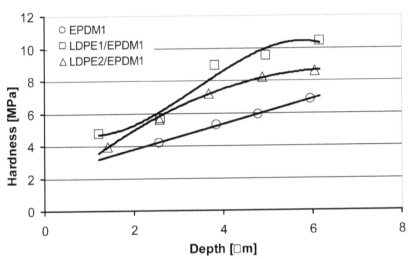

FIGURE 3.1   Hardness profile of LDPE/EPDM1 blends.

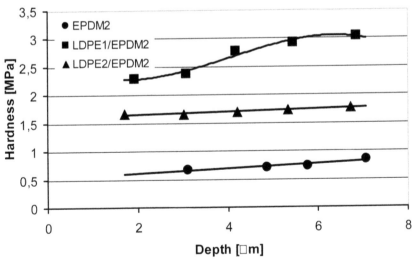

FIGURE 3.2   Hardness profile of LDPE/EPDM2 blends.

They confirm structural data. Blends of sequenced elastomer matrix (EPDM1) exhibit considerably higher hardness in comparison to the systems of amorphous one (EPDM2). It concerns both: bulk as well as the surface layer of materials. Improvement of mechanical properties of EPDM by blending with LDPE is easier to achieve when amorphous elastomer is to be modified. Linear polyethylene (LDPE1) is better than more branched one (LDPE2) in terms of reaching higher hardness. The difference is especially pronounced in bulk, gradually diminishing towards the surface of materials.

3.4   CONCLUSIONS

- EPDM matrix solvates crystalline phase of LDPE. The effect is significant for blends of amorphous elastomer matrix. In the case of sequenced EPDM matrix a part of crystalline phase of LDPE recrystallizes in less perfect form, which is not detectable by DSC.
- Improvement of mechanical properties of EPDM by blending with LDPE is higher for the elastomer of sequenced structure, confirming structural data. The best mechanical properties, obtained when linear plastomer is admixed to amorphous elastomer, stays in agreement with the highest degree of crystallinity of the systems studied.
- Degree of crystallinity of polymer components and their blends are higher in bulk than in the surface layer. Blends containing sequenced EPDM matrix exhibit the most significant difference, no matter molecular structure of LDPE, whereas the systems containing amorphous elastomer matrix facilitates recrystallization of linear polyethylene on the surface.
- Lamellas of crystalline phase of the surface layer of polyolefin blends studied are thicker than present in the surface layer of their components, what suggests cocrystallization of ethylene monomer unit from EPDM. Linear LDPE facilitates the phenomenon, especially when takes place in amorphous elastomer matrix. Branched plastomer recrystallizes to the same lamellar thickness, no matter the structure of elastomer matrix.
- Microindentation data reveals hardness profile of LDPE/EPDM blends, staying in agreement with structural data for the surface layer of systems studied.

KEYWORDS

- **Mechanical properties**
- **Polyolefine blends**
- **Structure**
- **Surface layer**

REFERENCES

1. Hamdani, S., Longuet, C., Lopez-Cuesta, J-M., & Ganachaud, F. (2010). Calcium and aluminum-based fillers as flame-retardant additives in silicone matrices. I. Blend preparation and thermal properties, Polym Degr Stabil, 95, 1911–1919.
2. Hamdani-Devarennes, S., Pommier, A., Longuet, C., Lopez-Cuesta, J-M., & Ganachaud, F. (2011). Calcium and aluminum-based fillers as flame-retardant additives in silicone matrices. II. Analyzes on composite residues from an industrial-based pyrolysis test, Polym Degr Stabil, 96, 1562–1572.
3. Hamdani-Devarennes, S., Longuet, C., Sonnier, R., Ganachaud, F., & Lopez-Cuesta, J-M. (2013). Calcium and aluminum-based fillers as flame-retardant additives in silicone matrices. III. Investigations on fire reaction, Polym Degr Stabil, 98, 2021–2032.
4. Mansouri, J., Wood, C. A., Roberts, K., Cheng, Y. B., & Burford, R. P. (2007). Investigation of the ceramifying process of modified silicone-silicate compositions, J Mater Sci, 42, 6046–6055.
5. Hanu, L. G., Simon, G. P., Mansouri, J., Burford, R. P., & Cheng, Y. B. (2004). Development of polymer-ceramic composites for improved fire resistance, J Mater Process Tech, 153–154, 401–407.
6. Bieliński, D. M., Anyszka, R., Pędzich, Z., & Dul, J. (2012). Ceramizable silicone rubber-based composites. Int J Adv Mater Manuf Charact, 1, 17–22.
7. Hanu, L. G., Simon, G. P., & Cheng, Y. B. (2006). Thermal stability and flammability of silicone polymer composites, Polym Degr Stabil, 91, 1373–1379.
8. Pędzich, Z., Bukanska, A., Bieliński, D. M., Anyszka, R., Dul, J., & Parys, G. (2012). Microstructure evolution of silicone rubber-based composites during ceramization at different conditions. Int J Adv Mater Manuf Charact, 1, 29–35.
9. Pędzich, Z., & Bieliński, D. M. (2010). Microstructure of silicone composites after ceramization Compos, 10, 249–254.
10. Dul, J., Parys, G., Pędzich, Z., Bieliński, D. M., & Anyszka, R. (2012). Mechanical properties of silicone-based composites destined for wire covers, Int J Adv Mater Manuf Charact, 1, 23–28.
11. Hamdani, S., Longuet, C., Perrin, D., Lopez-Cuesta, J-M., & Ganachaud, F. (2009). Flame retardancy of silicone-based materials, Polym Degr Stabil, 94, 465–495.
12. Mansouri, J., Burford, R. P., Cheng, Y. B., & Hanu, L. (2005). Formation of strong ceramified ash from silicone-based composites, J Mater Sci, 40, 5741–5749.

13. Mansouri, J., Burford, R. P., & Cheng, Y. B. (2006). Pyrolysis behavior of silicone-based ceramifying composites, Mater Sci Eng A, 425, 7–14.
14. Hanu, L. G., Simon, G. P., & Cheng, Y. B. (2005). Preferential orientation of muscovite in ceramifiable silicone composites, Mater Sci Eng A, 398, 180–187.
15. Xiong, Y., Shen, Q., Chen, F., Luo, G., Yu, K., & Zhang, L. (2012). High strength retention and dimensional stability of silicone/alumina composite panel under fire, Fire Mater, 36, 254–263.
16. Pędzich, Z., Anyszka, R., Bieliński, D. M., Ziąbka, M., Lach, R., & Zarzecka-Napierała, M. (2013). Silicon-basing ceramizable composites containing long fibers, J Mater Sci Chem Eng, 1, 43–48.
17. Bieliński, D. M., Anyszka, R., Pędzich, Z., Parys, G., & Dul, J. (2012). Ceramizable silicone rubber composites. Influence of type of mineral on ceramization, Compos, 12, 256–261.
18. Pędzich, Z., Bukańska, A., Bieliński, D. M., Anyszka, R., Dul, J., & Parys, G. (2012). Microstructure evolution of silicone rubber-based composites during ceramization in different conditions, Compos, 12, 251–255.
19. Morgan, A. B., Chu, L. L., & Harris, J. D. (2005). A flammability performance comparison between synthetic and natural clays in polystyrene nanocomposites, Fire Mater, 29, 213–229.
20. Wang, T., Shao, H., & Zhang, Q. (2010). Ceramifying fire-resistant polyethylene composites, Adv Compos Lett, 19, 175–179.
21. Shanks, R. A., Al-Hassany, Z., & Genovese, A. (2010). Fire-retardant and fire-barrier poly(vinyl acetate) composites for sealant application, Express Polym Lett, 4, 79–93.
22. Thomson, K. W., Rodrigo, P. D. D., Preston, C.M., & Griffin, G. J. (2006). Ceramifying polymers for advanced fire protection coatings, Proceedings of European Coatings Conference 2006, 15th September, Berlin, Germany.
23. Heiner, J., Stenberg, B., & Persson, M. (2003). Crosslinking of siloxane elastomers, Polym Test, 22, 253–257.
24. Ogunniyi, S. D. (1999). Peroxide vulcanization of rubber, Prog Rubber Plast Tech, 15, 95–112.
25. Anyszka, R., Bieliński, D. M., & Kowalczyk, M. (2013). Influence of dispersed phase selection on ceramizable silicone composites cross-linking, Elastom, 17, 16–20.
26. Bieliński, D., Ślusarski, L., Włochowicz, A., Ślusarczyk., & Cz. Douillard, A. (1997). Polimer Int., 44, 161.
27. Bieliński, D. Ślusarski, L. Włochowicz, A., & Douillard, A. (1997). Composite Interf., 5, 155.
28. Bieliński, D. Włochowicz, A. Dryzek, J., & Ślusarczyk, Cz. (2001). Composite Interf., 8, 1.
29. Bieliński, D., & Kaczmarek, Ł. (2006). J. Appl. Polym. Sci, 100, 625.
30. Barton, A. F. M. (1981). Handbook of Solubility Parameters and Other Cohesion Parameters, CRC Press, Boca Raton, FL.
31. Bojarski, J., & Lindeman, J. (1963). Polyethylene, 109, WNT, Warsaw.
32. Hindeleh, A. M., & Johnson, J. (1971). J. Phys. D: Appl. Phys. 4, 259.
33. Rosenbrock, H. H., & Storey, C. (1966)..Computational Techniques for Chemical Engineers, Pergamon Press, New York.
34. Vonk, C.G. (1970). J. Appl. Crystal., 8, 340.

35. Hendricks, R.W., & Schmidt, P. W. (1967). Acta Phys. Austriaca, 26, 97.
36. Hendricks, R.W., & Schmidt, P. W. (1970). Acta Phys. Austriaca, 37, 20.
37. Brandrup, J., & Immergut, E. H. (1989). Polymer Handbook 3rd Ed., Ch. 5, John Wiley & Sons, London-New York.

CHAPTER 4

A STUDY ON THE STRUCTURAL FEATURES OF PHB-EPC BLENDS AND THEIR THERMAL DEGRADATION

A. A. OL'KHOV, A. L. IORDANSKII, W. TYSZKIEWIC, and G. E. ZAIKOV

CONTENTS

ABSTRACT

The methods of DSC and IR spectroscopy were used to study various blends of poly(3-hydroxybu-tyrate) with ethylene-propylene copolymer rubber (EP). When the weight fractions of the initial polymers are equal, a phase inversion takes place; as the blends are enriched with EP, the degree of crystallinity of poly(3-hydroxybutyrate) decreases. In blends, the degradation of poly(3-hydroxybutyrate) begins at a lower temperature compared to the pure polymer and the thermooxidative activity of the ethylene-propylene copolymer in the blend decreases in comparison with the pure copolymer.

4.1 INTRODUCTION

Composite materials based on biodegradable polymers are currently evoking great scientific and practical interest. Among these polymers is poly(3-hydroxybu-tyrate) (PHB), which belongs to the class of poly(3-hydroxyalkanoates). Because of its good mechanical properties (close to those of PP) and biodegradability, PHB has been intensely studied in the literature [1]. However, because of its significant brittleness and high cost, PHB is virtually always employed in the form of blends with starch, cellulose, PE [2], etc., rather than in pure form.

This work is concerned with the study of the structural features of PHB-EPC blends and their thermal degradation.

4.2 EXPERIMENTAL PART

The materials used in this study were EPC of CO-059 grade (Dutral, Italy) containing 67.4 mol % ethylene units and 32.6 mol % propylene units. PHB with $M_{\eta} = 2.5 \, ' \, 105$ (Biomer, Germany) was used in the form of a fine powder. The PHB: EPC ratios were as follows: 100:0, 80:20, 70:30, 50:50, 30:70, 20:80, and 0:100 wt%.

The preliminary mixing of the components was performed using laboratory bending microrolls (brand VK-6) under heating: the microroll diameter was 80 mm, the friction coefficient was 1.4, the low-speed roller revolved at 8 rpm, and the gap between the rolls was 0.05 mm. The blending took place at 150 °C for 5 min.

Films were prepared by pressing using a manual heated press at 190 °C and at a pressure of 5 MPa; the cooling rate was ~50 °C/min.

The thermophysical characteristics of the tested films and the data on their thermal degradation were obtained using a DSM-2 M differential scanning calorimeter (the scanning rate was 16 K/min); the sample weight varied from 8 to 15 mg; and the device was calibrated using indium with Tm = 156.6 °C. To determine the degree of crystallinity, the melting heat of the crystalline PHB (90 J/g) was used [2]. The T_m and T_a values were determined with an accuracy up to 1 °C. The degree of crystallinity was calculated with an error up to ±10%. The structure of polymer chains was determined using IR spectroscopy (Specord M-80). The bands used for the analysis were structure-sensitive bands at 720 and 620 cm^{-1}, which belong to EPC and PHB, respectively [3]. The error in the determination of reduced band intensities did not exceed 15%.

4.3   RESULTS AND DISCUSSION

The melting endotherms of PHB, EPC, and their blends are shown in Fig. 4.1. Apparently, all the first melting thermograms (except for that of EPC) show a single peak characteristic of PHB.

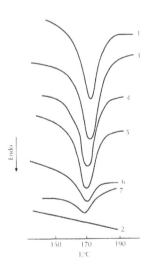

FIGURE 4.1   The melting endotherms of (1) PHB, (2) EPC, and their blends with compositions (3) 80: 20, (4) 70: 30, (5) 50: 50, (6) 30: 70, and (7) 20: 80 wt %.

The thermophysical characteristics obtained using DSC for blends of various compositions are listed in Table 4.1. As is apparent from this table, the melting heat rH_{ml} of PHB during first melting changes just slightly in comparison with the starting polymer. During cooling, only a single peak corresponding to the crystallizing PHB additionally appears.

TABLE 4.1   The Thermophysical Properties of PHB-EPC Films

PHB : EPC, wt %	T_m, °C	ΔH^*_{m1}	ΔH^*_{m2}	T_{cr}, °C	Degree of crystallini-ty**, %
		J/g			
100 : 0	174	88.3	88.9	64	98
80 : 20	173	75.8	76.2	64	84
70 : 30	172	59.5	60.1	60	66
50 : 50	172	56.4	52.1	62	63
30 : 70	172	29.3	20.5	60	33
20 : 80	171	22.3	15.7	–	25
0 : 100	–	–	–	–	–

* Calculated as areas under the melting curves: the first melting and the melting after the recrystallization, respectively.

** Calculated according to the ΔH^*_{m1} values.

However, the repeated melting endotherms of some blends (70% PHB + 30% EPC, 50% PHB + 50% EPC) display a low-temperature shoulder. Note that the melting enthalpy significantly changes as one passes from an EPC-enriched blend to a composition where PHB is predominant. When the content of EPC is high, the melting heat AHm2 of the recrystallized PHB significantly decreases. This effect should not be regarded as a consequence of the temperature factor, because the material was heated up to 195 °C during the DSM-2 M experiment and the films were prepared at 190 °C; the scanning rate was significantly lower than the cooling rate during the formation of the films (50 K/min). Thus, the state of the system after remelting during DSC measurements is close to equilibrium.

These results make it possible to assume that the melting heat and the degree of crystallinity of PHB decrease in EPC-enriched blends due to the mutual segmental solubility of the polymers [4] and due to the appearance of an extended interfacial layer. Also note that the degree of crystallinity may decrease because of the slow structural relaxation of the rigid-chain PHB. This, in turn, should affect the nature of interaction between the blend components. However, the absence of significant changes in the Tm and Ta values of PHB in blends indicates that EPC does not participate in nucleation during PHB crystallization and the decrease in the melting enthalpy of PHB is not associated with a decrease in the structural relaxation rate in its phase. Thus, the crystallinity of PHB decreases because of its significant amorphization related to the segmental solubility of blend components and to the presence of the extended interfacial layer.

Figure 4.2 shows the IR spectra for two blends of different compositions. As is known, the informative structure-sensitive band for PHB is that at 1228 cm^{-1} [5]. Unfortunately, the intensity of this band cannot be clearly determined in the present case, because it cannot be separated from the EPC structural band at 1242 cm^{-1} [3]. The bands used for this work were the band at 620 cm^{-1} (PHB) and the band at 720 cm^{-1} (EPC) [6], which correspond to vibrations of C-C bonds in methylene sequences (CH$_2$), where $n > 5$, occurring in the trans-zigzag conformation. The ratios between the optical densities of the bands at 720 and 620 cm^{-1} (D_{720}/D_{620}) are transformed in the coordinates of the equation where (5 is the fraction of EPC and W is the quantity characterizing a change in the ratio between structural elements corresponding to regular methylene sequences in EPC and PHB.

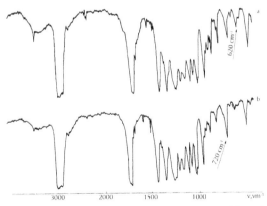

FIGURE 4.2   The IR spectra of PHB-EPC blends with compositions (a) 80:20 and (b) 20:80 wt %.

Figure 4.3 demonstrates the value of W plotted as a function of the blend composition. Apparently, this dependence is represented by a straight line in these coordinates but shows an inflection point. The latter provides evidence that phase inversion takes place and that the nature of intermolecular interactions between the polymer and the rubber changes.

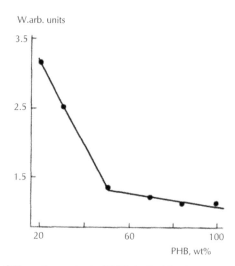

FIGURE 4.3   Plot of W vs. the content of PHB in the blend.

$$W = \log\left[D_{720}\beta / D_{620}\left(1-\beta\right)\right] + 2$$

where 5 is the fraction of EPC and W is the quantity characterizing a change in the ratio between structural elements corresponding to regular methylene sequences in EPC and PHB.

Figure 4.3 demonstrates the value of W plotted as a function of the blend composition. Apparently, this dependence is represented by a straight line in these coordinates but shows an inflection point. The latter provides evidence that phase inversion takes place and that the nature of intermolecular interactions between the polymer and the rubber changes.

The phase inversion causes the blends in question to behave in different ways during their thermal degradation. The DSM-2 M traces (Fig. 4.4) were measured in the range 100–500 °C. The thermograms of the blends display exothermic peaks of the thermal oxidation of EPC in the range

370–400 °C and endothermic peaks of the thermal degradation of PHB at $T > 250$ °C. For the pure PHB and EPC, the aforementioned peaks are observed in the ranges 200-300 °C and 360–430 °C, respectively. The blend samples studied in this work display two peaks each, thus confirming the existence of two phases. Note that the peak width increases (curves 3, and 4 in Fig. 4.4), and the heat Q of the thermal degradation of PHB changes in all the blends studied here (Table 4.2). This effect is apparently determined by the blend structure rather than by its composition. In blends, PHB becomes more active compared to the pure polymer and the rate of its thermal degradation increases. The temperature corresponding to the onset of thermal degradation 7^ decreases from 255 °C; the value characteristic of the pure PHB, to 180 °C (Table 4.2). The structure of the polymer becomes less perfect in this case; two likely reasons for this are a change in the morphology and the appearance of an extended interfacial layer.

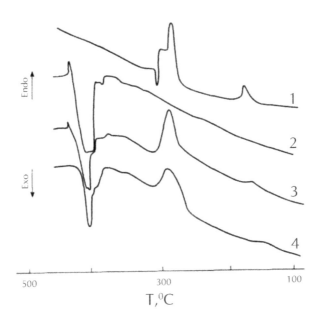

FIGURE 4.4   The DSC traces of (1) PHB, (2) EPC, and their blends with compositions (3) 70:30 and (4) 30:70 wt %.

TABLE 4.2   The Parameters of the Thermal Degradation Process

PHB : EPC, wt %	T_{od}(EPC). °C	T_{od}(PHB), °C	Q^*(PHB), kJ/g
100 : 0	–	255	0.53
70 : 30	370	180	1.38
30 : 70	380	250	0.51
0 : 100	360	--	--

* The specific heat of thermal degradation per g of PHB.

As to EPC, it acquires a higher thermal stability in the blends under examination, as indicated by the increase in the temperature corresponding to the onset of its thermal oxidation 7° (Table 4.2). The position of the exothermic peaks on the temperature scale characteristic of EPC indicates that its activity in blends is lower than that in the pure sample. The low-temperature shoulder of the exothermic EPC peak in the range 360–380 °C (Fig. 4.4) decreases with increasing content of PHB. Apparently, this effect is due to a change in the copolymer structure related to the interpenetration of PHB and EPC segments.

Thus, the existence of two peaks in DSC thermograms of the blends indicates the presence of two phases in the PHB-EPC blends. The phase inversion takes place in the vicinity of the composition with equal component weights. The components influence each other during film formation, and, hence, the appearance of the extended interfacial layer is presumed for samples containing more than 50% EPC. A change in the structure of the blends affects their thermal degradation. The degradation of PHB in blends is more pronounced than that in the pure PHB, but the thermal oxidation of EPC is retarded.

KEYWORDS

- **Degradation**
- **Degree of crystallinity**
- **Ethylene-propylene copolymer rubber**
- **Poly(3-hydroxybu-tyrate)**
- **Thermal stability**
- **Thermophysical properties**

REFERENCES

1. Seebach, D., Brunner, A., Bachmann, B. M., Hoffman, T., Kuhnle, F. N. M., & Len-gweier, U. D. (1996). *Biopolymers and Biooligomers of (R)-3-Hydroxyalkanoic Acids*: Contribution of Synthetic Organic Chemists, Zurich: Edgenos-sische Technicshe Hochschule.
2. Ol'khov, A. A., Vlasov, S. V., Shibryaeva, L. S., Litvi-nov, I. A., Tarasova, N. A., Kosenko, R. Yu., & lordan-skii, A. L. (2000). *Polymer Science, Ser. A, 42(4)*, 447.
3. Elliot, A. (1969). *Infra-Red Spectra and Structure of Organic Long-Chain Polymers*, London: Edward Arnold.
4. Lipatov, Yu. S. (1980). *Mezhfaznye yavleniya v polimerakh* (Interphase Phenomena in Polymers), Kiev: Naukova Dumka.
5. Labeek, G., Vorenkamp, E.J., & Schouten, A.J. (1995). *Macromolecules, 28(6)*, 2023.
6. Painter, P. C., Coleman, M. M., & Koenig, J. L. (1982). The Theory of Vibrational Spectroscopy and Its Application to Polymeric Materials, *New York: Wiley*.

PART 2

INTERPHASE ELASTOMER-FILLER INTERACTIONS

CHAPTER 5

THE STRUCTURE OF NANOFILLER IN ELASTOMERIC PARTICULATE-FILLED NANOCOMPOSITES

YU. G. YANOVSKII, G. V. KOZLOV, R. M. KOZLOWSKI, J. RICHERT, and G. E. ZAIKOV

CONTENTS

ABSTRACT

It has been shown that nanofiller particles (aggregates of particles) "chains" in elastomeric nanocomposites are physical fractal within the self-similarity (and, hence, fractality) range ~500–1450 nm. The low dimensions of nanofiller particles (aggregates of particles) structure in elastomeric nanocomposites are due to high fractal dimension of nanofiller initial particles surface.

5.1   INTRODUCTION

It is well known [1, 2] that in particulate-filled elastomeric nanocomposites (rubbers) nanofiller particles form linear spatial structures ("chains"). At the same time in polymer composites, filled with disperse microparticles (microcomposites) particles (aggregates of particles) of filler form a fractal network, which defines polymer matrix structure (analog of fractal lattice in computer simulation) [3]. This results to different mechanisms of polymer matrix structure formation in micro- and nanocomposites. If in the first filler particles (aggregates of particles) fractal network availability results to "disturbance" of polymer matrix structure, that is expressed in the increase of its fractal dimension d_f [3], then in case of polymer nanocomposites at nanofiller contents change the value d_f is not changed and equal to matrix polymer structure fractal dimension [4]. As it has to been expected, composites indicated classes structure formation mechanism change defines their properties change, in particular, reinforcement degree.

At present there are several methods of filler structure (distribution) determination in polymer matrix, both experimental [5, 6] and theoretical [3]. All the indicated methods describe this distribution by fractal dimension D_n of filler particles network. However, correct determination of any object fractal (Hausdorff) dimension includes three obligatory conditions. The first from them is the indicated above determination of fractal dimension numerical magnitude, which should not be equal to object topological dimension. As it is known [7], any real (physical) fractal possesses fractal properties within a certain scales range [8]. And at last, the third condition is the correct choice of measurement scales range itself. As it has been shown in papers [9, 10], the minimum range should exceed at any rate one self-similarity iteration.

The present paper purpose is dimension D_n estimation, both experimentally and theoretically, and checking two indicated above conditions fulfillment, i.e. obtaining of nanofiller particles (aggregates of particles) network ("chains") fractality strict proof in elastomeric nanocomposites on the example of particulate-filled butadiene-styrene rubber.

5.2   EXPERIMENTAL PART

The elastomeric particulate-filled nanocomposite on the basis of butadiene-styrene rubber (BSR) was an object of the study. The technical carbon of mark № 220 (TC) of industrial production, nano- and microshungite (the mean filler particles size makes up 20, 40 and 200 nm, accordingly) were used as a filler. All fillers content makes up 37 mass %. Nano- and microdimensional disperse shungite particles were obtained from industrially extractive material by processing according to the original technology. A size and polydispersity of the received in milling process shungite particles were monitored with the aid of analytical disk centrifuge (CPS Instruments, Inc., USA), allowing to determine with high precision the size and distribution by sizes within the range from 2 nm up to 50 mcm.

Nanostructure was studied on atomic-power microscopes Nano-DST (Pacific Nanotechnology, USA) and Easy Scan DFM (Nanosurf, Switzerland) by semi-contact method in the force modulation regime. Atomic-power microscopy results were processed with the aid of specialized software package SPIP (Scanning Probe Image Processor, Denmark). SPIP is a powerful programs package for processing of images, obtained on SPM, AFM, STM, scanning electron microscopes, transmission electron microscopes, interferometers, confocal microscopes, profilometers, optical microscopes and so on. The given package possesses the whole functions number, which are necessary at images precise analysis, in the number of which the following are included:

1.   the possibility of three-dimensional reflecting objects obtaining, distortions automatized leveling, including Z-error mistakes removal for examination separate elements and so on;

2.   quantitative analysis of particles or grains, more than 40 parameters can be calculated for each found particle or pore: area, perim-

eter, average diameter, the ratio of linear sizes of grain width to its height distance between grains, coordinates of grain center of mass a.a. can be presented in a diagram form or in a histogram form.

5.3   RESULTS AND DISCUSSION

The first method of dimension D_n experimental determination uses the following fractal relationship [11, 12]:

$$D_n = \frac{\ln N}{\ln \rho},\qquad(1)$$

where N is a number of particles with size ρ.

Particles sizes were established on the basis of atomic-power microscopy data (see Fig. 5.1).

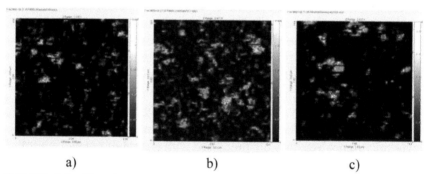

a)   b)   c)

FIGURE 5.1 The electron micrographs of nanocomposites BSR/TC (a), BSR/nanoshungite (b) and BSR/microshungite (c), obtained by atomic-power microscopy in the force modulation regime.

For each from the three studied nanocomposites no less than 200 particles were measured, the sizes of which were united into 10 groups and mean values N and ρ were obtained. The dependences $N(\rho)$ in double logarithmic coordinates were plotted, which proved to be linear and the values D_n were calculated according to their slope (see Fig. 5.2).

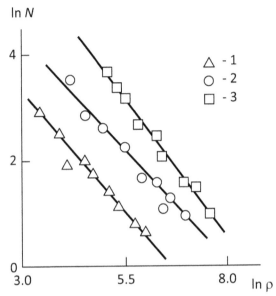

FIGURE 5.2 The dependence of nanofiller particles number N on their size r for nanocomposites BSR/TC (1), BSR/nanoshungite (2) and BSR/microshungite (3).

It is obvious, that at such approach fractal dimension D_n is determined in two-dimensional Euclidean space, whereas real nanocomposite should be considered in three-dimensional Euclidean space. The following relationship can be used for D_n re-calculation for the case of three-dimensional space [13]:

$$D3 = \frac{d + D2 \pm \left[(d - D2)^2 - 2\right]^{1/2}}{2}, \qquad (2)$$

where $D3$ and $D2$ are corresponding fractal dimensions in three- and two-dimensional Euclidean spaces, $d=3$.

The calculated according to the indicated method dimensions D_n are adduced in Table 5.1. As it follows from the data of Table 5.1, the values D_n for the studied nanocomposites are varied within the range of 1.10–1.36, that is, they characterize more or less branched linear formations ("chains") of nanofiller particles (aggregates of particles) in elastomeric nanocomposite structure. Let us remind that for particulate-filled composites polyhydroxiether/graphite the value D_n changes within the range of approx. 2.30–2.80 [5], that is, for these materials filler particles network is a bulk object, but not a linear one [7].

TABLE 5.1   The Dimensions of Nanofiller Particles (aggregates of particles) Structure in Elastomeric Nanocomposites

The nanocomposite	D_n, the equations (1)	D_n, the equations (3)	d_0	d_{surf}	φ_n	D_n, the equations (7)
BSR/TC	1.19	1.17	2.86	2.64	0.48	1.11
BSR/nanoshungite	1.10	1.10	2.81	2.56	0.36	0.78
BSR/microshungite	1.36	1.39	2.41	2.39	0.32	1.47

Another method of D_n experimental determination uses the so-called "quadrates method" [14]. Its essence consists in the following. On the enlarged nanocomposite microphotograph (see Fig. 5.1) a net of quadrates with quadrate side size α_i, changing from 4.5 up to 24 mm with constant ratio $\alpha_{i+1}/\alpha_i = 1.5$, is applied and then quadrates number N_i, in to which nanofiller particles hit (fully or partly), is calculated. Five arbitrary net positions concerning microphotograph were chosen for each measurement. If nanofiller particles network is fractal, then the following relationship should be fulfilled [14]:

$$N_i \sim S_i^{-D_n/2}, \tag{3}$$

where S_i is quadrate area, which is equal to α_i^2.

In Fig. 5.3, the dependences of N_i on S_i in double logarithmic coordinates for the three studied nanocomposites, corresponding to the Eq. (3), is adduced. As one can see, these dependences are linear, that allows to determine the value D_n from their slope. The determined according to the Eq. (3) values D_n are also adduced in Table 5.1, from which a good correspondence of dimensions D_n, obtained by the two described above methods, follows (their average discrepancy makes up 2.1% after these dimensions re-calculation for three dimensional space according to the Eq. (2)).

As it has been shown in Ref. [15], at the Eq. (3) the usage for self-similar fractal objects the condition should be fulfilled:

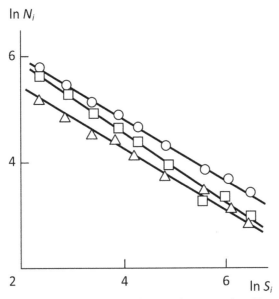

FIGURE 5.3 The dependences of covering quadrates number N_i on their area S_i, corresponding to the relationship (3), in double logarithmic coordinates for nanocomposites on the basis of BSR. The designations are the same, that in Fig. 5.2.

$$N_i - N_{i-1} \sim S_i^{-D_n}. \qquad (4)$$

In Fig. 5.4 the dependence, corresponding to the Eq. (4), for the three studied elastomeric nanocomposites is adduced. As one can see, this dependence is linear, passes through coordinates origin, that according to the Eq. (4) is confirmed by nanofiller particles (aggregates of particles) "chains" self-similarity within the selected α_i range. It is obvious, that this self-similarity will be a statistical one [15]. Let us note, that the points, corresponding to $\alpha_i=16$ mm for nanocomposites BSR/TC and BSR/microshungite, do not correspond to a common straight line. Accounting for electron microphotographs of Fig. 5.1 enlargement this gives the self-similarity range for nanofiller "chains" of 464–1472 nm. For nanocomposite BSR/nanoshungite, which has no points deviating from a straight line of Fig. 5.4, α_i range makes up 311–1510 nm, that corresponds well enough to the indicated above self-similarity range.

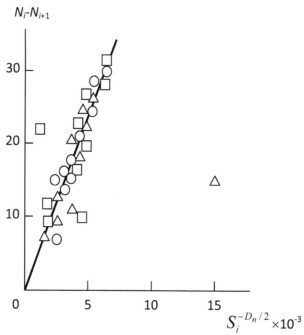

FIGURE 5.4 The dependences of $(N_i - N_{i+1})$ on the value $S_i^{-D_n/2}$, corresponding to the relationship (4), for nanocomposites on the basis of BSR. The designations are the same, that in Fig. 5.2.

In Refs. [9, 10] it has been shown, that measurement scales S_i minimum range should contained at least one self-similarity iteration. In this case the condition for ratio of maximum S_{max} and minimum S_{min} areas of covering quadrates should be fulfilled [10]:

$$\frac{S_{max}}{S_{min}} > 2^{2/D_n}.$$

(5)

Hence, accounting for the defined above restriction let us obtain $S_{max}/S_{min} = 121/20.25 = 5.975$, that is larger than values $2^{2/D_n}$ for the studied nanocomposites, which are equal to 2.71–3.52. This means, that measurement scales range is chosen correctly.

The self-similarity iterations number μ can be estimated from the inequality [10]:

$$\left(\frac{S_{max}}{S_{min}}\right)^{D_n/2} > 2^{\mu} \tag{6}$$

Using the indicated above values of the included in the inequality Eq. (6) parameters, $\mu=1.42–1.75$ is obtained for the studied nanocomposites, that is, in our experiment conditions self-similarity iterations number is larger than unity, that again is confirmed by the value D_n estimation correctness [6].

And let us consider in conclusion the physical grounds of smaller values D_n for elastomeric nanocomposites in comparison with polymer microcomposites, i.e. the causes of nanofiller particles (aggregates of particles) "chains" formation in the first. The value D_n can be determined theoretically according to the equation [3]:

$$\varphi_{if} = \frac{D_n + 2.55d_0 - 7.10}{4.18}, \tag{7}$$

where φ_{if} is interfacial regions relative fraction, d_0 is nanofiller initial particles surface dimension.

The dimension d_0 estimation can be carried out with the aid of the relationship [4]:

$$S_u = 410\left(\frac{D_p}{2}\right)^{d_0-d}, \tag{8}$$

where S_u is nanofiller initial particles specific surface in m^2/g, D_p is their diameter in nm, d is dimension of Euclidean space, in which a fractal is considered (it is obvious, in our case $d=3$).

The value S_u can be calculated according to the equation [16]:

$$S_u = \frac{6}{\rho_n D_p}, \tag{9}$$

where ρ_n is nanofiller density, which is determined according to the empirical formula [4]:

$$\rho_n = 0.188(D_p)^{1/3} \tag{10}$$

The results of value d_0 theoretical estimation are adduced in Table 5.1. The value φ_{if} can be calculated according to the equation [4]:

$$\varphi_{if} = \varphi_n(d_{surf} - 2), \tag{11}$$

where φ_n is nanofiller volume fraction, d_{surf} is fractal dimension of nanoparticles aggregate surface.

The value φ_n is determined according to the equation [4]:

$$\varphi_n = \frac{W_n}{\rho_n}, \tag{12}$$

where W_n is nanofiller mass fraction and dimension d_{surf} is calculated according to the Eqs. (8)–(10) at diameter D_p replacement on nanoparticles aggregate diameter D_{agr}, which is determined experimentally (see Fig. 5.5).

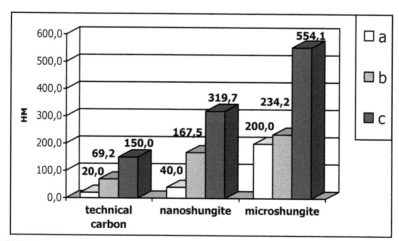

FIGURE 5.5   The initial particles diameter (a), their aggregates size in nanocomposite (b) and distance between nanoparticles aggregates (c) for nanocomposites on the basis of BSR, filled with technical carbon, nano- and microshungite.

The results of dimension D_n theoretical calculation according to the Eqs. (7)–(12) are adduced in Table 5.1, from which theory and experiment good correspondence follows. The Eq. (7) indicates unequivocally the cause of filler in nano- and microcomposites different behavior. The high (close to 3, see Table 5.1) values d_0 for nanoparticles and relatively small (d_0=2.17 for graphite) values d_0 for microparticles at comparable values φ_{if} for composites of the indicated classes [3, 4].

5.4   CONCLUSIONS

Therefore, the present paper results have shown, that nanofiller particles (aggregates of particles) "chains" in elastomeric nanocomposites are physical fractal within self-similarity (and, hence, fractality [12]) range of ~ 500–1450 nm. In this range their dimension D_n can be estimated according to the Eqs. (1), (3) and (7). The cited examples demonstrate the necessity of the measurement scales range correct choice. As it has been noted earlier [17], linearity of the plots, corresponding to the Eqs. (1) and (3), and D_n nonintegral value do not guarantee object self-similarity (and, hence, fractality). The nanofiller particles (aggregates of particles) structure low dimensions are due to the initial nanofiller particles surface high fractal dimension.

KEYWORDS

- "Chains"
- Fractal analysis
- Nanocomposite
- Nanofiller
- Structure

REFERENCES

1. Lipatov, Yu S. (1977). *The Physical Chemistry of Filled Polymers*. Moscow, Khimiya, 304 p.
2. Bartenev, G. M., Zelenev, Yu V. (1983). *The Physics and Mechanics of Polymers*. Moscow, Vysshaya Shkola, 391 p.

3. Kozlov, G. V., Yanovskii, Yu G., & Zaikov, G. E. (2010). *Structure and Properties of Particulate-Filled Polymer Composites: the Fractal Analysis.* New York, Nova Science Publishers, Inc., 282 p.

4. Mikitaev, A. K., Kozlov, G. V., & Zaikov, G. E. (2008). *Polymer Nanocomposites: the Variety of Structural Forms and Applications.* New York, Nova Science Publishers, Inc., 319 p.

5. Kozlov, G. V., & Mikitaev, A. K. (1996). *Mekhanika Kompozitsionnykh Materialov i Konstruktsii, 2(3–4)*, 144–157.

6. Kozlov, G. V., Yanovskii, Yu G., & Mikitaev, A. K. (1998). *Mekhanika Kompozitnykh Materialov, 34(4)*, 539–544.

7. Balankin, A. S. (1991). *Synergetics of Deformable Body.* Moscow, Publishers of Ministry Defence SSSR, 404 p.

8. Hornbogen, E. (1989). *Intern. Mater. Rev, 34(6)*, 277–296.

9. Pfeifer, P. (1984). *Appl. Surf. Sci., 18(1)*, 146–164.

10. Avnir, D., Farin, D., & Pfeifer, P. J. (1985). *Colloid Interface* Sci, *103(1)*, 112–123.

11. Ishikawa, K. J. (1990). *Mater. Sci. Lett, 9(4)*, 400–402.

12. Ivanova, V. S., Balankin, A. S., Bunin, I. Zh., & Oksogoev, A. A. (1994). *Synergetics and Fractals in Material Science.* Moscow, Nauka, 383 p.

13. Vstovskii, G. V., Kolmakov, L. G., & Terent'ev, V. F. (1993). *Metally, (4)*, 164–178.

14. Hansen, J. P., & Skjeitorp, A. T. (1988). *Phys. Rev. B, 38(4)*, 2635–2638.

15. Pfeifer, P., Avnir, D., & Farin, D. J. (1984). *Stat. Phys, 36(5/6)*, 699–716.

16. Bobryshev, A. N., Kozomazov, V. N., Babin, L. O., & Solomatov, V. I. (1994). Synergetics of Composite Materials. *Lipetsk, NPO ORIUS*, 154 p.

17. Farin, D., Peleg, S., Yavin, D., & Avnir, D. (1985). *Langmuir, 1(4)*, 399–407.

CHAPTER 6

THE DESCRIPTION OF NANOFILLER PARTICLES AGGREGATION WITHIN THE FRAMEWORKS OF IRREVERSIBLE AGGREGATION MODELS

YU. G. YANOVSKII, G. V. KOZLOV, K. PYRZYNSKI, and G. E. ZAIKOV

CONTENTS

ABSTRACT

A nanofiller disperse particles aggregation process in elastomeric matrix has been studied. The modified model of irreversible aggregation particle-cluster was used for this process theoretical analysis. The modification necessity is defined by simultaneous formation of a large number of nanoparticles aggregates. The offered approach allows to predict nanoparticles aggregates final parameters as a function of the initial particles size, their contents and other factors number.

6.1   INTRODUCTION

The aggregation of the initial nanofiller powder particles in more or less large particles aggregates always occurs in the course of technological process of making particulate-filled polymer composites in general [1] and elastomeric nanocomposites in particular [2]. The aggregation process tells on composites (nanocomposites) macroscopic properties [1, 3]. For nanocomposites nanofiller aggregation process gains special significance, since its intensity can be the one, that nanofiller particles aggregates size exceeds 100 nm – the value, which assumes (although and conditionally enough [4]) as an upper dimensional limit for nanoparticle. In other words, the aggregation process can result to the situation, when primordially supposed nanocomposite ceases to be the one. Therefore at present several methods exist, which allowed to suppress nanoparticles aggregation process [2, 5]. Proceeding from this, in the present paper theoretical treatment of disperses nanofiller aggregation process in butadiene-styrene rubber matrix within the frameworks of irreversible aggregation models was carried out.

6.2   EXPERIMENTAL PART

The elastomeric particulate-filled nanocomposite on the basis of butadiene-styrene rubber was an object of the study. Mineral shungite nanodimensional and microdimensional particles and also industrially produced technical carbon with mass contents of 37 mass % were used as a filler. The analysis of the received in milling process shungite particles were monitored with the aid of analytical disk centrifuge (CPS Instruments,

Inc., USA), allowing to determine with high precision the size and distribution by sizes within the range from 2 nm up to 50 mcm.

Nanostructure was studied on atomic-power microscopes Nano-DST (Pacific Nanotechnology, USA) and Easy Scan DFM (Nanosurf, Switzerland) by semi-contact method in the force modulation regime. Atomic-power microscopy results were processed with the aid of specialized software package SPIP (Scanning Probe Image Processor, Denmark). SPIP is a powerful programs package for processing of images, obtained on scanning probe microscopy (SPM), atomic forced microscopy (AFM), scanning tunneling microscopy (STM), scanning electron microscopes, transmission electron microscopes, interferometers, confocal microscopes, profilometers, optical microscopes and so on. The given package possesses the whole functions number, which are necessary at images precise analysis, in the number of which the following are included:

1. the possibility of three-dimensional reflected objects obtaining, distortions automatized leveling, including Z-error mistakes removal for examination separate elements and so on;
2. quantitative analysis of particles or grains, more than 40 parameters can be calculated for each found particle or pore: area, perimeter, average diameter, the ratio of linear sizes of grain width to its height distance between grains, coordinates of grain center of mass a.a. can be presented in a diagram form or in a histogram form.

6.3 RESULTS AND DISCUSSION

For theoretical treatment of nanofiller particles aggregate growth processes and final sizes traditional irreversible aggregation models are inapplicable, since it is obvious, that in nanocomposites aggregates a large number of simultaneous growth takes place. Therefore the model of multiple growth, offered in Ref. [6], was used for nanofiller aggregation description.

In Fig. 6.1, the images of the studied nanocomposites, obtained in the force modulation regime, and corresponding to them nanoparticles aggregates fractal dimension d_f distributions are adduced. As it follows from the adduced values d_f (d_f=2.40–2.48), nanofiller particles aggregates in the studied nanocomposites are formed by a mechanism particle-cluster (P-Cl), that is, they are Witten-Sander clusters [7]. The variant A, was chosen which according to mobile particles are added to the lattice, consisting of a large number of "seeds" with density of c_0 at simulation beginning [6]. Such model generates structures, which have fractal geometry on length short scales with value $d_f \gg 2.5$ (see Fig. 6.1) and homogeneous structure on length large scales. A relatively high particles concentration c is required in the model for uninterrupted network formation [6].

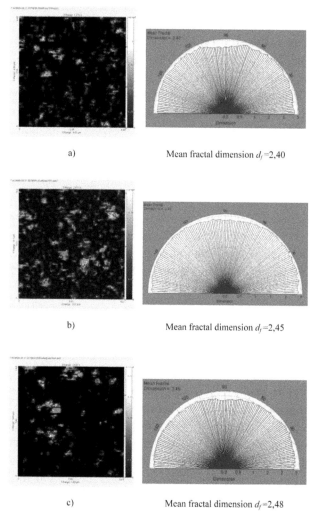

a)   Mean fractal dimension $d_f = 2,40$

b)   Mean fractal dimension $d_f = 2,45$

c)   Mean fractal dimension $d_f = 2,48$

FIGURE 6.1 The images, obtained in the force modulation regime, for nanocomposites, filled with technical carbon (a), nanoshungite (b), microshungite (c) and corresponding to them fractal dimensions d_f.

In case of "seeds" high concentration c_0 for the variant A the following relationship was obtained [6]:

$$R_{max}^{d_f} = N = c/c_0,$$ (1)

where R_{max} is nanoparticles cluster (aggregate) greatest radius, N is nanoparticles number per one aggregate, c is nanoparticles concentration, c_0 is "seeds" number, which is equal to nanoparticles clusters (aggregates) number.

The value N can be estimated according to the following equation [8]:

$$2R_{max} = \left(\frac{S_n N}{\pi \eta}\right)^{1/2},$$ (2)

where S_n is cross-sectional area of nanoparticles, from which aggregate consists, η is packing coefficient, equal to 0.74.

The experimentally obtained nanoparticles aggregate diameter $2R_{agr}$ was accepted as $2R_{max}$ (Table 6.1) and the value S_n was also calculated according to the experimental values of nanoparticles radius r_n (Table 6.1). In Table 6.1 the values N for the studied nanofillers, obtained according to the indicated method, were adduced. It is significant that the value N is a maximum one for nanoshungite despite larger values r_n in comparison with technical carbon.

TABLE 6.1 The Parameters of Irreversible Aggregation Model of Nanofiller Particles Aggregates Growth

Filler	Experimental radius of nanofiller aggregate R_{agr}, nm	Radius of nanofiller particle r_n, nm	Number of particles in one aggregate N	Radius of nanofiller aggregate R_{max}^T, the Eq. (1), nm	Radius of nanofiller aggregate R_{agr}^T, the Eq. (3), nm	Radius of nanofiller aggregate R_c, the Eq. (8), nm
Technical carbon	34.6	10	35.4	34.7	34.7	33.9
Nano-shungite	83.6	20	51.8	45.0	90.0	71.0
Micro-shungite	117.1	100	4.1	15.8	158.0	255.0

Further the Eq. (1) allows to estimate the greatest radius R_{max}^T of nanoparticles aggregate within the frameworks of the aggregation model [6]. These values R_{max}^T are adduced in Table 6.1, from which their reduction in a sequence of technical carbon-nanoshungite-microshungite that fully contradicts to the experimental data, that is, to R_{agr} change (Table 6.1). However, we must not neglect the fact that the Eq. (1) was obtained within the frameworks of computer simulation, where the initial aggregating particles sizes are the same in all cases [6]. For real nanocomposites the values r_n can be distinguished essentially (Table 6.1). It is expected, that the value R_{agr} or R_{max}^T will be the higher, the larger is the radius of nanoparticles, forming aggregate, is i.e. r_n. Then theoretical value of nanofiller particles cluster (aggregate) radius R_{agr}^T can be determined as follows:

$$R_{agr}^T = k_n r_n N^{1/d_f} , \qquad (3)$$

where k_n is proportionality coefficient, in the present work accepted empirically equal to 0.9.

The comparison of experimental R_{agr} and calculated according to the equation (3) R_{agr}^T values of the studied nanofillers particles aggregates radius shows their good correspondence (the average discrepancy of R_{agr} and R_{agr}^T makes up 11.4 %). Therefore, the theoretical model [6] gives a good correspondence to the experiment only in case of consideration of aggregating particles real characteristics and, in the first place, their size.

Let us consider two more important aspects of nanofiller particles aggregation within the frameworks of the model [6]. Some features of the indicated process are defined by nanoparticles diffusion at nanocomposites processing. Specifically, length scale, connected with diffusible nanoparticle, is correlation length ξ of diffusion. By definition, the growth phenomena in sites, remote more than ξ, are statistically independent. Such definition allows to connect the value ξ with the mean distance between nanofiller particles aggregates L_n. The value ξ can be calculated according to the equation [6]:

$$\xi^2 \approx \tilde{n}^{-1} R_{agr}^{d_f - d + 2} , \qquad (4)$$

where c is nanoparticles concentration, d is dimension of Euclidean space, in which a fractal is considered (it is obvious, that in our case $d=3$).

The value c should be accepted equal to nanofiller volume contents φ_n, which is calculated as follows [9]:

$$\varphi_n = \frac{W_n}{\rho_n},$$ (5)

where W_n is nanofiller mass contents, ρ_n is its density, determined according to the equation [3]:

$$\rho_n = 0.188(2r_n)^{1/3},$$ (6)

The values r_n and R_{agr} were obtained experimentally (see histogram of Fig. 6.2). In Fig. 6.3, the relation between L_n and ξ is adduced, which, as it is expected, proves to be linear and passing through coordinates origin. This means, that the distance between nanofiller particles aggregates is limited by mean displacement of statistical walks, by which nanoparticles are simulated. The relationship between L_n and ξ can be expressed analytically as follows:

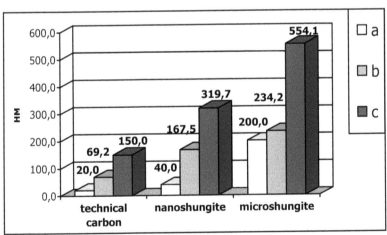

FIGURE 6.2 The initial particles diameter (a), their aggregates size in nanocomposite (b) and distance between nanoparticles aggregates (c) for nanocomposites, filled with technical carbon, nano- and microshungite.

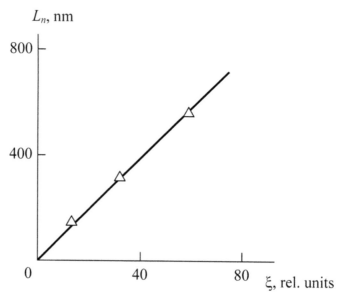

FIGURE 6.3   The relation between diffusion correlation length x and distance between nanoparticles aggregates L_n for studied nanocomposites.

$$L_n = 9.6\xi, \text{nm}. \tag{7}$$

The second important aspect of the model [6] in reference to nanofiller particles aggregation simulation is a finite nonzero initial particles concentration c or φ_n effect, which takes place in any real systems. This effect is realized at the condition $\xi \approx R_{agr}$, that occurs at the critical value $R_{agr}(R_c)$, determined according to the relationship [6]:

$$c \sim R_c^{d_f - d}. \tag{8}$$

The Eq. (8) right side represents cluster (particles aggregate) mean density. This equation establishes that fractal growth continues only, until cluster density reduces up to medium density, in which it grows. The calculated according to the Eq. (8) values R_c for the considered nanoparticles are adduced in Table 6.1, from which it follows, that they give reasonable correspondence with this parameter experimental values R_{agr} (the average discrepancy of R_c and R_{agr} makes up 24%).

Since the treatment [6] was obtained within the frameworks of a more general model of diffusion-limited aggregation, then its correspondence to the experimental data indicated unequivocally, that aggregation processes in these systems were controlled by diffusion. Therefore let us consider briefly nanofiller particles diffusion. Statistical walkers diffusion constant ζ can be determined with the aid of the relationship [6]:

$$\xi \approx (\zeta t)^{1/2}, \tag{9}$$

where t is walk duration.

The Eq. (9) supposes (at t=const) ζ increase in a number technical carbon-nanoshungite-microshungite as 196–1069–3434 relative units, that is, diffusion intensification at diffusible particles size growth. At the same time diffusivity D for these particles can be described by the well-known Einstein's relationship [10]:

$$D = \frac{kT}{6\pi\eta r_n \alpha}, \tag{10}$$

where k is Boltzmann constant, T is temperature, η is medium viscosity, α is numerical coefficient, which further is accepted equal to 1.

In its turn, the value η can be estimated according to the equation [11]:

$$\frac{\eta}{\eta_0} = 1 + \frac{2.5\varphi_n}{1 - \varphi_n}, \tag{11}$$

where η_0 and η are initial polymer and its mixture with nanofiller viscosity, accordingly, φ_n is nanofiller volume contents.

The calculation according to the Eqs. (10) and (11) shows, that within the indicated above nanofillers number the value D changes as 1.32–1.14–0.44 relative units, i.e. reduces in three times, that was expected. This apparent contradiction is due to the choice of the condition t=const (where t is nanocomposite production duration) in the equation (9). In real conditions the value t is restricted by nanoparticle contact with growing aggregate and then instead of t the value t/c_0 should be used, where c_0 is seeds concentration, determined according to the Eq. (1). In this case the value ζ for the indicated nanofillers changes as 0.288–0.118–0.086, that is, it reduces in 3.3 times, that corresponds fully to the calculation according to the Einstein's relationship (the Eq. (10)). This means, that nanoparticles diffusion in polymer matrix obeys classical laws of Newtonian rheology [10].

6.4 CONCLUSIONS

Disperse nanofiller particles aggregation in elastomeric matrix can be described theoretically within the frameworks of a modified model of irreversible aggregation particle-cluster. The obligatory consideration of nanofiller initial particles size is a feature of the indicated model application to real systems description. The indicated particles diffusion in polymer matrix obeys classical laws of Newtonian liquids hydrodynamics. The offered approach allows to predict nanoparticles aggregates final parameters as a function of the initial particles size, their contents and other factors number.

KEYWORDS

- **Aggregation**
- **Diffusion**
- **Elastomer**
- **Nanocomposite**
- **Nanoparticle**

REFERENCES

1. Kozlov, G. V., Yanovskii, Yu G., & Zaikov, G. E. (2010). Structure and Properties of Particulate-Filled Polymer Composites: the Fractal Analysis. New York, *Nova Science Publishers*, Inc, 282 p.
2. Edwards, D. C. (1990). Polymer-filler interactions in rubber reinforcement. *J. Mater. Sci, 25(12)*, 4175.
3. Mikitaev, A. K., Kozlov, G. V., & Zaikov, G. E. (2008). Polymer Nanocomposites: the Variety of Structural Forms and Applications. New York, *Nova Science Publishers*, Inc, 319 p.
4. Buchachenko, A. L. (2003). The nanochemistry – direct way to high technologies of new century. *Uspekhi Khimii, 72(5)*, 419.
5. Kozlov, G. V., Yanovskii, Yu G., Burya, A. I., & Aphashagova, Z. Kh. (2007). Structure and properties of particulate-filled nanocomposites phenylone/aerosol. *Mekhanika Kompozitsionnykh Materialov i Konstruktsii, 13(4)*, 479.
6. Witten, T. A., & Meakin, P. (1983). Diffusion-limited aggregation at multiple growth sites. *Phys. Rev. B, 28(10)*, 5632.

7. Witten, T. A., & Sander, L. M. (1983). Diffusion-limited aggregation. *Phys. Rev. B, 27(9)*, 5686.
8. Bobryshev, A. N., Kozomazov, V. N., Babin, L. O., & Solomatov, V. I. (1994). Synergetics of Composite Materials. *Lipetsk, NPO ORIUS,* 154 p.
9. Sheng, N., Boyce, M. C., Parks, D. M., Rutledge, G. C., Abes, J. I., & Cohen, R. E. (2004). Multiscale micromechanical modeling of polymer/clay nanocomposites and the effective clay particle. *Polymer, 45(2)*, 487.
10. Happel, J., & Brenner, G. (1976). The Hydrodynamics at Small Reynolds Numbers. *Moscow, Mir,* 418 p.
11. Mills, N. J. (1971). The rheology of filled polymers. *J. Appl. Polymer Sci, 15(11)*, 2791.

CHAPTER 7

THE INTERFACIAL REGIONS FORMATION MECHANISM IN ELASTOMERIC PARTICULATE-FILLED NANOCOMPOSITES

YU. G. YANOVSKII, G. V. KOZLOV, KH. SH. YAKH'YAEVA, Z. WERTEJUK, and G. E. ZAIKOV

CONTENTS

ABSTRACT

It has been shown that interfacial regions in particulate-filled elastomeric nanocomposites represent the polymer layer, adsorbed by nanofiller surface. This layer is formed by a diffusive mechanism. The last is realized at the expense of polymer matrix molecular mobility.

7.1   INTRODUCTION

As it is well known [1] that the interfacial interaction role in multiphase systems, including polymer composites, is very great. In polymer composites such interactions (interfacial adhesion) absence results in sharp reduction of their reinforcement degree [2]. For polymer nanocomposites interfacial adhesion existence in the first place means the formation of interfacial regions, which are the same reinforcing element for these materials, as nanofiller actually [3]. Proceeding from the said above, it is necessary to know the conditions and mechanisms of interfacial regions formation in polymer nanocomposites for their structure control. The present paper purpose is these mechanism definition and the indicated research is performed on the example of three particulate-filled nanocomposites on the basis of butadiene-styrene rubber.

7.2   EXPERIMENTAL PART

The elastomeric particulate-filled nanocomposites on the basis of butadiene-styrene rubber (BSR) was an object of the study. The technical carbon of mark N220 (TC) of industrial production, nano- and microshungite (the mean filler particles size makes up 20, 40 and 200 nm, accordingly) were used as a filler. All fillers content makes up 37 mass%. Nano- and micro-dimensional disperse shungite particles were prepared from industrially extractive material by processing according to the original technology. A size and polydispersity of the received in milling process shungite particles were monitored with the aid of analytical disk centrifuge (CPS Instruments, Inc., USA), allowing to determine with high precision the size and distribution by sizes within the range from 2 nm up to 50 mcm.

Nanostructure was studied on atomic-power microscopes Nano- DST (Pacific Nanotechnology, USA) and Easy DFM (Nanosurf, Switzerland) by semi-contact method in the force modulation regime. Atomic-power microscopy results were processed with the aid of specialized software package SPIP (Scanning Probe Image Processor, Denmark). SPIP is a powerful programs package for processing of images, obtained on SPM, AFM, STM, scanning electron microscopes, transmission electron microscopes, interferometers, confocal microscopes, optical microscopes and so on. The given package possesses the whole functions number, which are necessary at images precise analysis, the most important of which include:

1. the possibility of three-dimensional reflecting objects obtaining, distortions automatized leveling, including Z-error mistakes removal for the examination of separate elements and so on;
2. quantitative analysis of particles or grains, more than 40 parameters can be calculated for each found particle or pore: area, perimeter, average diameter, the ratio of grain width to its height, distance between grains, coordinates of grain center of mass a.a. can be presented in a diagram form or in a histogram form.

The tests by nanocomposite BSR/nanoshungite nanomechanical properties determination were carried out by nanoindentation method [4] on apparatus NanoTest 600 (Micro Materials, Great Britain) in loads wide range from 0.01 mN up to 2.0 mN. Sample indentation was conducted in 10 points with an interval of 30 mcm. The load was increased with constant rate up to the greatest given load reaching (for the rate 0.05 mN/s – 1 mN/s). The indentation rate was changed in conformity with the greatest load value counting, that loading cycle should take 20 s. The unloading was conducted with the same rate as the loading. In the given experiment the "Berkovich's indentor" was used with an angle at tip of 65.3° and rounding radius of 200 nm. Indentations were carried out in the checked load regime with preload of 0.001 mN.

For elasticity modulus calculation the obtained in the experiment by nanoindentation course dependences of load on indentation depth (strain) in ten points for each sample at loads of 0.01, 0.02, 0.03, 0.05, 0.10, 0.50, 1.0 and 2.0 mN were processed according to Oliver-Pharr method [5].

7.3 RESULTS AND DISCUSSION

In Ref. [6] at sorption phenomena on fractal objects analysis the following relationship was proposed for determination of relative fraction of the adsorbed phase on fractal objects, which can be considered as total fraction of nanofiller and interfacial (adsorbed) layer ϑ in case of polymer nanocomposites [7]:

$$\vartheta = \rho_n^{-1} m_0^{(1-3/D)},\tag{1}$$

where ρ_n is the density of a separate particle (in the considered case – nanofiller initial particle), m_0 is the mass of a separate particle (the indicated nanofiller particle also), D is object fractal dimension.

The value ρ_n can be estimated according to the empirical formula [3]:

$$\rho_n = 188 (D_p)^{1/3}, \text{ kg/m}^3,\tag{2}$$

where D_p is nanofiller initial particles diameter in nm (see Fig. 7.1) and the value m_0 was calculated in supposition of these particles spherical shape and the known geometrical relationships.

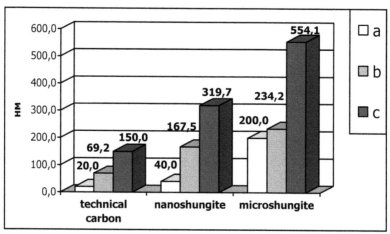

FIGURE 7.1 The initial particles diameter (a), their aggregates size in nanocomposite (b) and distance between particles in nanocomposite (c) for samples of BSR, filled with technical carbon, nano- and microshungite.

The dimension D identification requires certain discussion. In Ref. [6] it was implied that fractal dimension of nanofiller particles aggregate structure should be accepted as D. However, according to Refs. [8, 9] fractal dimension of the adsorbing object surface d_{surf}, but not dimension of the object itself, exercises decisive influence on the adsorbed (interfacial) layer formation. Let us note that nanofiller particles aggregate should be considered as an adsorbing object in the nanocomposites case and later on calculations $D=d_{surf}$ will be accepted.

Proceeding from the stated above considerations, it should be supposed, that the estimated according to the Eq. (1) value ϑ should be equal to sum $(\varphi_n + \varphi_{if})$, where φ_n and φ_{if} are relative volume fractions of nanofiller and interfacial regions. These parameters were calculated according to the equations [3]:

$$\varphi_n = \frac{W_n}{\rho_{ag}},$$ (3)

$$\varphi_{if} = \varphi_n \left(d_{surf} - 1 \right),$$ (4)

where W_n is nanofiller mass content, r_{ag} is nanofiller particles aggregate density, determined according to the Eq. (2) at the condition: $r_{ag} = r_n$ and $D_{ag} = D_p$, where D_{ag} is nanofiller particles aggregate diameter (see Fig. 7.1).

The dimension d_{surf} value was determined with the aid of the following equation [3]:

$$S_u = 410 \left(\frac{D_{ag}}{2} \right)^{d_{surf} - d},$$ (5)

where S_u is specific surface of nanofiller particles aggregate in m^2/g, d is dimension of Euclidean space, in which a fractal is considered (it is obvious, that in our case $d=3$).

The value S_u was calculated according to the equation [10]:

$$S_u = \frac{6}{\rho_{ag} D_{ag}},$$ (6)

In Fig. 7.2, the comparison of the parameters ϑ and $(\varphi_n + \varphi_{if})$, calculated according to the Eqs. (1), (3) and (4), is adduced. As one can see, between them the linear correlation, passing through coordinates origin, was obtained. This means that the indicated parameters coincide to within constant and are connected between themselves by the following analytical relationship:

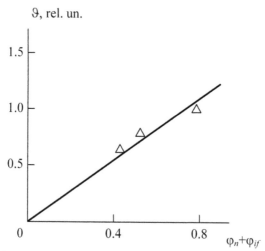

FIGURE 7.2   The dependence of adsorbed phase relative fraction J on sum of nanofiller and interfacial regions relative fractions (j_n+j_{if}) for nanocomposites on the basis of BSR.

$$\left(\varphi_n + \varphi_{if}\right) = 0.7149 .$$ (7)

The Eq. (1) allows the value ϑ estimation as a function of d_{surf} at fixed ρ_n and m_0, i.e. as a matter of fact a relative fraction of nanocomposite structure reinforcing element, since these materials reinforcement degree can be described by the equation [3]:

$$\frac{E_n}{E_m} = 1 + 11\left(\varphi_n + \varphi_{if}\right)^{1.7},$$ (8)

where E_n and E_m are elasticity moduli of nanocomposite and matrix polymer, respectively.

In Fig. 7.3 the dependence of ϑ (in relative units) on d_{surf} is adduced. As one can see, at d_{surf}=0–1 (i.e., for zero- and one-dimensional objects) the value ϑ is close to zero and according to the Eq. (8) this means reinforcement absence. Within the range of d_{surf}=1–2, that is, for nanofiller with porous surface, relatively small increasing ϑ is observed. And at last, at d_{surf}=2–3, that is, for nanoparticles with well developed surface, the most strong ϑ growth is obtained. In other words, the most attractive nanofiller will be the same, for which the surface dimension d_{surf} is higher than its topological dimension d=2. The data of Fig. 7.3 also explain the well-known dependence $E_n/E_m(D_p)$, where at D_p reduction up to ~ 10 nm the strong growth of E_n/E_m is observed and at $D_p<10$ nm – some sharper decrease of elastomeric nanocomposites reinforcement degree [11]. It is obvious, that from the practical point of view nanofiller particles with diameter smaller than 10 nm become point (zero-dimensional) objects that explains sharp reduction of E_n/E_m within the framework of the Eq. (1).

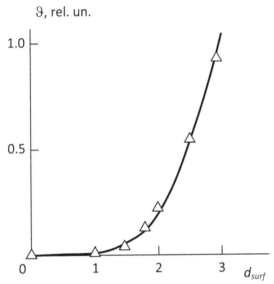

FIGURE 7.3 The dependence of adsorbed phase relative fraction J on nanofiller particles surface dimension d_{surf}.

Let us consider further diffusive processes influence on interfacial regions formation in the studied nanocomposites. In Refs. [12, 13], the treatment of depositions structure formation on fibers and surfaces within the frameworks of irreversible aggregation models was proposed. Within the framework of this treatment the relationship between mean-square deposition (interfacial layer) thickness l_{if} and particles (statistical segments) number n_i in it at $n_i \to \infty$ was proposed [13]:

$$l_{if} \sim n_i^{\varepsilon},\qquad(9)$$

where the exponent ε characterizes deposition (interfacial layer) formation mechanism: ε=1.7 for depositions, controlled by diffusion, and ε=1.0 – for the conditions, where diffusive processes are unimportant.

The value l_{if} was calculated according to the following fractal equation [14]:

$$l_{if} = a\left(\frac{D_{ag}}{2a}\right)^{2(d-d_{surf})/d}, \qquad (10)$$

where a is the lower scale of polymer matrix fractal behavior, which is accepted equal to statistical segment length l_{st}.

In its turn, the statistical segments number in interfacial layer n_i can be calculated according to the equation [14]:

$$n_i = \frac{\varphi_{if}}{Sl_{st}}, \qquad (11)$$

where S is cross-sectional area of macromolecule for polymer matrix.

Let us consider the determination of molecular characteristics S and l_{st} for butadiene-styrene rubber. As it is known [15], the value of macromolecule diameter square is equal to: for polybutadiene – 20.7 Å^2 and for polystyrene – 69.8 Å^2. Calculating the macromolecule, simulated as cylinder, cross-sectional area for the indicated polymers according to the known geometrical formulae, let us obtain 16.2 Å^2 and 54.8 Å^2, respectively. Further, accepting as S for butadiene-styrene rubber mean value of the cited above areas, let us obtain S=35.5 Å^2. Further the characteristic ratio C_{∞} can be determined, which is a polymer chain statistical flexibility indicator [16], with the aid of the following empirical formula [14]:

$$T_g = 129\left(\frac{S}{C_{\infty}}\right)^{1/2}, \qquad (12)$$

where T_g is glass transition temperature, which is equal to 217 K for BSR [3]. Let us obtain from the equation (12) C_{∞}=12.5 for BSR.

And at last, l_{st} value is determined according to the equation [17]:

$$l_{st} = l_0 C_\infty ,$$ (13)

where l_0 is the main chain skeletal bond length, which is equal to 0.154 nm [18].

In Fig. 7.4 the dependence $l_{if}(n_i)$ in double logarithmic coordinates is adduced, which turned out to be linear, that allows to determine according to its slope the exponent ε, which is equal to 1.7. This magnitude ε supposes, that interfacial layer in the studied nanocomposites is formed by a diffusive mechanism.

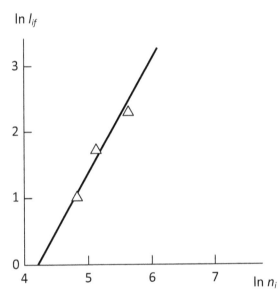

FIGURE 7.4   The dependence of interfacial layer thickness l_{if} on statistical segments number n_i in it for nanocomposites on the basis of BSR in double logarithmic coordinates.

Variation ε defines corresponding change of interfacial layer structure characterized by its fractal dimension d_f^{if} as well. Between parameters ε, d and d_f^{if} intercommunication exists, expressed by the relationship [13]:

$$\varepsilon = \frac{1}{d_f^{if} - d + 1} .$$ (14)

From the Eq. (14) it follows, that d_f^{if} =2.568. The indicated value d_f^{if} allows to calculate the value C_∞ (C_∞^{if}) for interfacial regions [14]:

$$C_\infty^{if} = \frac{2d_f^{if}}{d(d-1)(d-d_f^{if})} + \frac{4}{3}.$$ (15)

The value C_∞^{if} for interfacial regions (d_f^{if} =2.568) is equal to 3.31, that is, in interfacial regions macromolecular coils strong compactization occurs in comparison with volume polymer matrix of BSR (C_∞=12.5) [16]. Naturally, such molecular characteristics change should influence on polymeric material properties. In Fig. 7.5, the obtained according to the original methods results of elasticity moduli calculation for nanocomposite BSR/nanoshungite components, received in interpolation process of nanoindentation data and processed in program SPIP, are presented. These data (Fig. 7.5, on the right) showed that interfacial layers elasticity modulus was only by 23–45% lower than nanofiller elasticity modulus, but it is higher than the corresponding parameter of polymer matrix in 6.0–8.5 times. The adduced experimental data confirm, that for the studied nanocomposites interfacial layer is a reinforcing element to the same extent, as actually nanofiller (see the Eq. (8)).

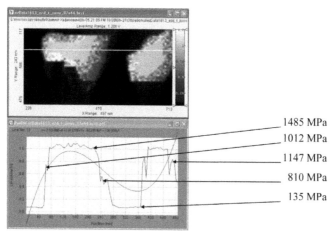

1485 MPa
1012 MPa
1147 MPa
810 MPa
135 MPa

FIGURE 7.5   The processed by SPIP image of nanocomposite BSR/nanoshungite, obtained by force modulation method, and mechanical characteristics of structural components according to the data of nanoindentation (strain 150 nm).

Let us fulfill several estimations, confirming self-congruence of the obtained results. For nanocomposites the relative fraction of densely-packed regions φ_{dens}, that is, $(\varphi_n + \varphi_{if})$, can be determined with the aid of the equation [3]:

$$d_f = 3 - 6 \times 10^{-10} \left(\frac{\varphi_{dens}}{SC_\infty} \right)^{1/2}, \tag{16}$$

where S is given in Å².

Calculation according to the Eq. (16) at $d_f = d_f^{if}$ and the indicated above values S and C_∞ gives $\varphi_{dens} = 0.609$ and the mean value $(\varphi_n + \varphi_{if})$ for the three studied nanocomposites is equal to 0.582, that is a good enough correspondence.

As it is known [14], the value of excess energy localization regions dimension D_f is equal approximately to C_∞. In its turn, the interfacial regions elasticity modulus E_{if} can be estimated according to the equation [19]:

$$E_{if} = 0.44 + 0.94(D_f - 3), \text{ GPa.} \tag{17}$$

Using the mean experimental value $E_{if} = 0.987$ GPa (see Fig. 7.5), let us obtain $D_f = 3.58$ from the Eq. (17), that is close again to the obtained above value $D_f = C_\infty^{if} = 3.31$.

The equation, obtained within the frameworks of solid mediums mechanics (continuous approach), is one of the methods of Poisson's ratio ν value estimation for particulate-filled composites [20]:

$$\frac{1}{\nu} = \frac{\varphi_n}{\nu_n} + \frac{1 - \varphi_n}{\nu_m}, \tag{18}$$

where ν, ν_n and ν_m are Poisson's ratios of composite, filler and polymer matrix, respectively.

Further the structure fractal dimension d_f can be determined according to the equation [21]:

$$d_f = (d - 1)(1 + \nu). \tag{19}$$

The mixtures rule using is another mode of the value d_f estimation [3]:

$$d_f = d_f^{if}\varphi_{if} + d_f^{n}\varphi_n + d_f^{m}\left(1 - \varphi_n - \varphi_{if}\right), \tag{20}$$

where d_f^{if}, d_f^{n} and d_f^{m} are fractal dimensions of interfacial regions, nanofiller and volume polymer matrix structure, which are accepted equal to 2.568, 2.50 and 2.95 [21], accordingly.

Calculation according to the Eqs. (19) and (20) lead to the following sets of d_f values: 2.653, 2.703, 2.766 and 2.617, 2.728 and 2.762 for nanocomposites BSR/TC, BSR/nanoshungite and BSR/microshungite, respectively. As one can see, both indicated estimation d_f methods give self-congruent results with mean discrepancy smaller than 1%.

And in conclusion of this problem let us note, that the calculated according to the equation (10) and determined experimentally (Fig. 7.5) l_{if} values are equal to 7.85 and 8.70 nm, respectively, for nanocomposite BSR/nanoshungite that again is a good enough correspondence.

7.4   CONCLUSIONS

Thus, the present paper results have shown that the interfacial regions in particulate-filled elastomeric nanocomposites represent the polymer layer, adsorbed by nanofiller surface, which is formed by diffusive mechanism. It is obvious, that the last one is realized at the expense of polymer matrix molecular mobility. Nanofiller particles (aggregates of particles) surface dimension influences strongly on the adsorbed layer formation process and, hence, on nanocomposites reinforcement degree. A performed estimations number gave self-congruent results with both experimental data and with calculated characteristics of other theoretical models that allows to use them for detailed quantitative description of structure and properties of both interfacial regions and nanocomposites as a whole.

KEYWORDS

- **Diffusive mechanism**
- **Interfacial layer**
- **Molecular mobility**
- **Nanocomposite**
- **Particles surface**

REFERENCES

1. Lipatov, Yu. S. (1980). Interfacial Phenomena in Polymers. *Naukova Dumka, Kiev*, 260.
2. Knunyants, N. N., Lyapunova, M. A., Manevich, L. I., Oshmyan, V. G., & Shaulov, A. Y. (1986). *Mekhanika Kompozithykh Materialov*, *22(2)*, 231–234.
3. Miktaev, A. K., Kozlov, G. V., & Zaikov, G. E. (2008). *Polymer Nanocomposites: Variety of Structural Forms and Applications.* Nova Science Publishers, Inc., New York, 319.
4. Kornev, Yu. V., Yumashev, O. B., Zhogin, V. A., Karnet, Yu N., Yanovskii, Yu G., & Gamlitskii, Yu A. (2008). *Kautschuk i Rezina*, 6, 18–23.
5. Oliver, W. C., & Pharr, G. M. (1992). *J. Mater. Res.*, *7(6)*, 1564–1583.
6. Yakubov, T. S. (1988). *Doklady AN SSSR* 303*(2)*, 425–428.
7. Kozlov, G. V., Aphashagova, Z. Kh., & Zaikov, G. E. (2009). *Vestnik MITKhT*, *4(3)*, 89–91.
8. Pfeifer, P. (1984). *Appl. Surf. Sci.*, *18(1)*, 146–164.
9. Pfeifer, P., Pietronero, In. L., and Tosatti, E. (1986). (Eds.), *Fractal in Physics*, Elsevier, Amsterdam, 70–79.
10. Bobryshev, A. N., Kozomazov, V. N., Babin, L. O., & Solomatov, V. I. (1994). Synergetics of Composite Materials. *NPO ORIUS, Lipetsk*, 154.
11. Edwards, D. C. (1990). *J. Mater. Sci.*, *25(12)*, 4175–4185.
12. Meakin, P. (1983). *Phys. Rev. A*, *27(5)*, 2616–2623.
13. Meakin, P. (1984). *Phys. Rev. B*, *30(8)*, 4207–4214.
14. Kozlov, G. V., Yanovskii, Yu G., & Zaikov, G. E. (2010). Structure and Properties of Particulate-Filled Polymer Composites: the Fractal Analysis. *Nova Science Publishers*, Inc., New York, 282.
15. Aharoni, S. M. (1985)*Macromolecules*, *18(12)*, 2624–2630.
16. Budtov, V. P. (1992). *Physical Chemistry of Polymer Solutions. Khimiya, Sankt-Peterburg*, 384.
17. Wu, S. (1989). *J. Polymer Sci.: Part B: Polymer Phys.*, *27(4)*, 723–741.
18. Aharoni, S. M. (1983). *Macromolecules*, *16(9)*, 1722–1728.
19. Miktaev, A. K., & Kozlov, G. V. (2008). The Fractal Mechanics of Polymeric Materials. Publishers of *KBSU, Nalchik*, 312.

20. Kubat, J., Rigdahl, M., & Welander, M. (1990). *J. Appl. Polymer Sci.*, *39(5)*, 1527–1539.
21. Balankin, A. S. (1991). Synergetics of Deformable Body. Publishers of *Ministry Defence of SSSR*, Moscow, 404.

CHAPTER 8

BORON OXIDE AS A FLUXING AGENT FOR SILICONE RUBBER-BASED CERAMIZABLE COMPOSITES

R. ANYSZKA, D. M. BIELIŃSKI, and Z. PĘDZICH

CONTENTS

ABSTRACT

Boron oxide particles were incorporated to silicone rubber-based mixes containing fumed silica (reinforcing filler) and reference mineral fillers – aluminum hydroxide, wollastonite, calcined kaolin, mica (phlogipite) and surface modified montmorillonite with dimethyl-dihydrogenatedtallow quaternary ammonium salt. Acidic character of boron oxide, which can disturb the peroxide curing process, was compensated by addition of magnesium oxide. The influence of boron oxide particles on properties of composites was determined and mechanism of their ceramization process studied.

Vulcanization kinetic was studied. Mechanical properties of cured composites before and after various regimes of ceramization were investigated. Ceramic phase obtained after heat treatment was characterized by porosimetry and scanning electron microscopy. Morphology and mechanical properties of composites before and after ceramization depend on the type of mineral filler applied, whereas the kinetic of vulcanization is different only for composite containing surface modified montmorillonite.

8.1 INTRODUCTION

Ceramizable (ceramifiable) silicone rubber-based composites are fire resistant materials developed especially for cable covers application. In case of fire, electrical installations are endangered of short circuit effect which can deactivate lots of important devices, like fire sprinklers, elevators, fire alarms or lamps indicating route to emergency exits. Ceramizable composites are able to sustain functioning of electric circuit on fire and high temperature up to 120 min by producing ceramic, porous layer protecting copper wire inside a cable.

Ceramizable silicone-based materials are dispersion type of composites, in which mineral particles (refractory fillers and, in some compositions, fluxing agent particles) are dispersed in continuous phase of silicone rubber [1–18]. Mechanism of protecting ceramic shield creation on the border between fire and material includes:

1. Production of amorphous silica during thermal degradation of polysiloxane matrix under oxidizing atmosphere, which results in

creation of mineral bridges between refractory filler particles (Fig. 8.1).

2. Sintering of mineral filler particles due to condensation of hydroxyl groups present on their surface (Fig. 8.2) [19].

3. Production of new mineral phases as a result of reaction between primary filler particles and silica produced as a result of thermal degradation of silicone matrix (Fig. 8.3).

4. Creation of strong glassy bridges between particles of high thermal resistive minerals by low temperature softening amorphous glass particles (Fig. 8.4). This type of ceramization made possible to use hydrocarbon polymers as a continuous phase for ceramizable composites. In the subject literature it is possible to find papers describing ceramizable composites based on polyethylene [20], poly(vinyl acetate) [21], poly(vinyl chloride) and ethylene-propylene-diene rubber (EPDM) [22].

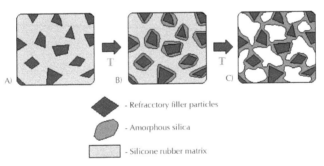

FIGURE 8.1   Scheme of ceramization process based on creation of amorphous silica microbridges between refractory filler particles during thermo oxidizing degradation of silicone rubber matrix.

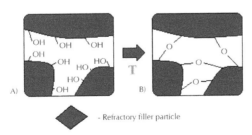

FIGURE 8.2   Scheme of sintering process of refractory mineral particles involving hydroxyl groups condensation.

$$PDMS \longrightarrow SiO2 + cyclosiloxanes \qquad (>500 \,^{\circ}C)$$

$$CaCO_3 \longrightarrow CaO + CO_2 \qquad\qquad (>600 \,^{\circ}C)$$

$$CaO + SiO_2 \longrightarrow CaSiO_3 \text{ (Wollastonite) } (800 \,^{\circ}C)$$

$$2CaO + SiO_2 \longrightarrow Ca_2SiO_4 \text{ (Lamite)} \qquad (800 \,^{\circ}C)$$

FIGURE 8.3 Possible chemical reactions between components of ceramizable composites, taking place at temperature.

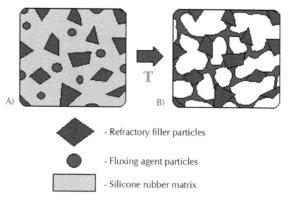

A)   B)

◆ - Refractory filler particles

● - Fluxing agent particles

▭ - Silicone rubber matrix

FIGURE 8.4 Scheme of the ceramization process based on creation of microbridges between refractory filler particles by melted particles of fluxing agent.

In this chapter, particles of boron oxide were used as a fluxing agent in order to enhance mechanical strength of ceramic layer after high temperature treatment of silicone rubber-based composites. Due to relatively low meting point temperature of crystalline boron oxide ($T_m = 450\,°C$), which is comparable, or even lower than the temperature of softening point of glass oxide frits, very effective ceramization process, resulting in desirable thermal and mechanical properties of the composites, is expected.

8.2   EXPERIMENTAL PART

8.2.1   MATERIALS

Silicone rubber (HTV) containing 0.07% of vinyl groups, produced by "Polish Silicones" Ltd. (Poland), reinforced by "Aerosil 200" fumed silica produced by "Evonik Industries" (Germany), an elastomer base. Boron oxide from "Alfa Aesar GmbH & Co KG" (Germany) we used as a fluxing agent, together with magnesium oxide "MagChem 50" obtained from "Martin Marietta Magnesia Specialties" (USA), applied for decreasing acidic character of the composition. Calcined kaolin "Polestar 200R" originated from "Imerys Minerals" Ltd (France), mica (phlogopite) "PW 150" from "Minelco Minerals" (Sweden), wollastonite "Termin 939–304" from "Quarzwerke Gruppe" (Germany), aluminum hydroxide "Martinal OL-107 LEO" from "Martinswerk" (Germany) and surface modified montmorillonite "Cloisite 20A" using dimethyl-dihydrogenatedtallow quaternary ammonium salt, produced by "Rockwood Additives Ltd." (UK) were applied as refractory filler. 50% paste of 2,4-dichlorobenzoyl peroxide originated from "Novichem" (Poland) was used for crosslinking of the silicone rubber composites.

8.2.2   SAMPLE PREPARATION

Due to quite large size of boron oxide particles, the first step was to grind them in a Pulverisette 5 planetary mill made by "Fritsch GmbH" (Germany) using bowls covered internally with agate (280 cm^3 of capacity) and 20 mm of diameter agate balls. The mill operated with the speed of 350 rpm, during 30 min (Fig. 8.5).

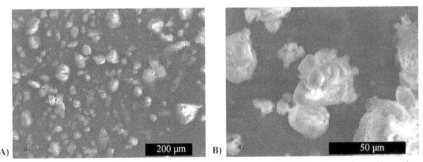

FIGURE 8.5   Boron oxide particles after grinding process, SEM photographs made under magnification of (a) 150× and (b) 1000×.

Studied mixes containing boron oxide and other mineral fillers (Table 8.1) were prepared with a Brabender-Plasticorder laboratory mixer (Germany), whose rotors were operating with 20 rpm during components incorporation (5 min) and with 60 rpm during their homogenization (10 min). Composite samples were vulcanized in a steel mold, at 130 °C during time t_{90} determined rheometrically according to ISO 3417.

TABLE 8.1   Composition of the Ceramizable Composites Studied

Composites Components	C-KAO	WOL	ALOH	M-MMT	MIC
Silicone rubber	100	100	100	100	100
Fumed silica	50	50	50	50	50
B_2O_3	20	20	20	20	20
MgO	5	5	5	5	5
Calcined kaolin	10	–	–	–	–
Wollastonite	–	10	–	–	–
Aluminum hydroxide	–	–	10	–	–
Surface modified montmoril-lonite	–	–	–	10	–
Mica (phlogopite)	–	–	–	–	10
2,4-dichlorobenzoyl peroxide (50% paste)	1.8	1.8	1.8	1.8	1.8

8.2.3   SAMPLE CERAMIZATION

Composite samples were ceramized in a laboratory furnace under four different regimes of heat treatment. Isothermal heating at 600, 800 or 1000 °C during 20 min, and slow temperature speed ratio from room temperature up to 1000 °C during 2 h. Then they were cooled down in open air prior to testing.

8.2.4   TECHNIQUES

Vulcanization kinetics of the composite mixes were determined using a WG-02 vulcameter produced by Metalchem (Poland). Mechanical properties of the composites: elongation at break (EB), mechanical moduli at 100, 200 and 300% of elongation (SE_{100}, SE_{200}, SE_{300}) tensile strength (TS)

and tear strength (TES) were measured with a Zwick-Roell 1435 instrument (Germany). Hardness of vulcanized composites was determined using a Zwick-Roell tester (Germany), applying force of 12,5 N.

Composites after heat treatment were tested for porosity, using a mercury porosimeter CarloErba 2000 (Italy) and endurance under compression using a Zwick Roell Z 2.5 tester (Germany). Compression tests were made on the samples of cylindrical geometry (Fig. 8.6).

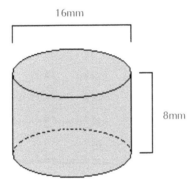

FIGURE 8.6   Cylindrical shape of samples ceramized in laboratory furnace.

For complementary microstructure analysis of the ceramized composite samples, a Hitachi S-4700 (Japan) scanning electron microscope equipped with a BSE detector was used

8.3   RESULTS AND DISCUSSION

8.3.1   VULCANIZATION KINETICS OF COMPOSITES

Vulcanization kinetics of the composites vulcanization are very similar regardless refractory filler type, excluding the mix containing surface modified montmorillonite (M-MMT) (Table 8.2). In its case values of torque were significantly lower than for other samples: minimum (M_{min}), optimum (M_{opt}) and maximum (M_{max}) torque values, respectively, from 14 to 27%, from 34 to 40% and from 35 to 41%. This effect can by explained by plasticization of silicone matrix by quaternary ammonium salt present in M-MMT. Also values of scorch (t_{02}) and vulcanization (t_{90}) time were higher in the case of M-MMT sample: 47 to 127% and from 30 to 50% than other mixes studied for the scorch time and vulcanization time, respectively. This fact could cause problems, because of high-speed extrusion technology commonly used in cable industry.

TABLE 8.2 Vulcameter's Parameter of the Composite Mixes Studied. Minimum (M_{min}), Optimum (M_{opt}) and Maximum (M_{max}) Torque Value, Scorch Time (t_{02}) and Vulcanization Time (t_{90})

Composites Properties	C-KAO	WOL	ALOH	M-MMT	MIC
M_{min} [dNm]	59.7	61.3	62.3	45.5	52.8
M_{opt} [dNm]	136.2	140.0	145.5	87.1	132.8
M_{max} [dNm]	144.7	148.8	154.7	91.7	141.7
t_{02} [s]	29	34	22	50	34
t_{90} [s]	82	92	95	123	92

8.3.2 MECHANICAL PROPERTIES OF COMPOSITES BEFORE CERAMIZATION

Values of stress at 100% (SE_{100}), 200% (SE_{200}), 300% (SE_{300}) of elongation, tensile strength (TS) and elongation at break (EB) of the composites are very similar, except M-MMT sample filled with surface modified montmorillonite (Table 8.3). In its case values of moduli at 100, 200 and 300% of elongation and tensile strength were slightly lower than for other composites, but its elongation at break was almost two times higher than determined for the other samples. Also M-MMT sample characterize itself by the best tear strength higher from 19% up to even 90% than composites filled with unmodified refractory powders. The highest value of hardness has the composite filled with wollastonite (WOL) whereas the lowest the composite containing surface modified montmorillonite (M-MMT). However, the difference is not significant – it does not exceed ca 10%.

TABLE 8.3 Mechanical Properties of the Composites Studied

Composites Properties	C-KAO	WOL	ALOH	M-MMT	MIC
SE_{100} [MPa]	1.9	1.7	1.7	1.3	2.0
SE_{200} [MPa]	2.8	2.6	2.6	1.5	2.7
SE_{300} [MPa]	3.7	3.5	3.4	1.8	3.4
TS [MPa]	3.9	3.9	3.7	3.1	3.9
EB [%]	322	350	338	641	357
TES [N/mm]	8.2	10.4	9.3	15.6	13.1
H [°ShA]	63.1	68.0	66.2	62.4	64.1

Values of Stress at 100% (SE_{100}), 200% (SE_{200}) and 300% (SE_{300}) of Composites Elongation and their Tensile Strength (TS), Tear Strength (TES), Elongation at Break (EB) and Hardness (H).

8.3.3 MECHANICAL PROPERTIES OF COMPOSITES AFTER HEAT TREATMENT

The best mechanical properties, after isothermal heating exhibit composite sample containing mica (phlogopite) (MIC) (Table 8.4). The weakest ceramic phases were obtained after isothermal heating of the composites at 800 °C, whereas the strongest structures were created after slow heating from room temperature to 1000 °C. After this treatment the sample containing surface modified montmorillonite (M-MMT) and filled with mica flakes (MIC) exhibits the best mechanical strength (ca. 600 N).

All of the composites shrunk after slow heating treatment, except sample containing mica (MIC), which expanded in diameter by 2 mm. The largest shrinkage was detected for the samples filled with aluminum hydroxide (ALOH) and surface modified montmorillonite (M-MMT), which shrunk in diameter by 5 mm. Changes to shape of composites subjected to the action of high temperature is very important from the point of view of maintaining continuity of ceramic cover protecting copper wire of a cable on fire (Fig. 8.7).

TABLE 8.4   Force required to Break Samples Ceramized under Various Conditions and Changes to the Sample Diameter after Slow Ceramization (heating from room temperature to 1000 °C during 2 h)

Composites Properties	C-KAO	WOL	ALOH	M-MMT	MIC
600 °C 20 min. [N]	20.0	34.4	33.8	17.8	53.9
800 °C 20 min. [N]	16.2	27.3	17.4	17.2	28.0
1000 °C 20 min. [N]	20.2	67.3	22.1	35.8	109.0
$T_R \rightarrow 1000$ °C [N]	478	548	505	613	601
Sample parameters					
Diameter of samples after $T_R \rightarrow 1000$ °C heating [mm]	13	15	11	11	18
Changes of samples diameter after $T_R \rightarrow 1000$ °C heating [mm]	–3	–1	–5	–5	+2

C-KAO   WOL   ALOH M-MMT   MIC

FIGURE 8.7   Photographs of the composite samples subjected to slow heating from room temperature to 1000 °C during 2 h.

8.3.4   MORPHOLOGY OF COMPOSITES AFTER HEAT TREATMENT

Scanning electron microscopy (SEM) pictures of composite samples ceramized slowly under heating from room temperature to 1000 °C during 2 h demonstrate good adhesion between ceramic phase components in the composites filled with mica (MIC – C1, C2), wollastonite (WOL – E1, E2) and aluminum hydroxide (ALOH – A1, A2). Far worse adhesion is observed for the samples containing calcined kaolin (C-KAO – B1, B2) and surface modified montmorillonite (M-MMT – D1, D2) (Fig. 8.8).

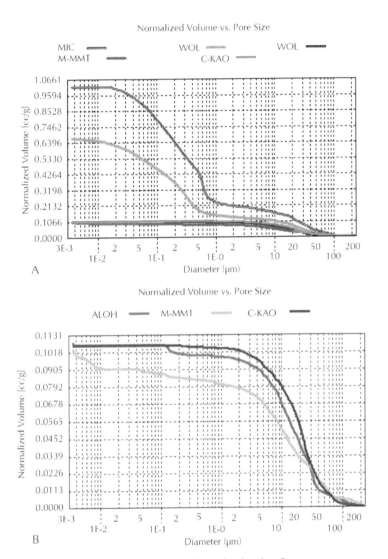

FIGURE 8.8 Micromorphology of composites after heating form room temperature to 1000 °C during 2 h SEM photographs of composites containing (a) aluminum hydroxide, (b) calcined kaolin, (c) mica, (d) surface modified montmorillonite and (e) wollastonite.

Porosimetry analysis demonstrates that samples containing mica and wollastonite have created nano-porous structure with high amount of pores (Fig. 8.9a). This kind of structure is expected to be the best from the point of view of mechanical properties as well as thermal insulation of the material. Amount of pores in samples filled with calcined kaolin (C-KAO), aluminum hydroxide (ALOH) and surface treated montmoril-lonie (M-MMT) is almost ten times lower in comparison to the composite containing mica (MIC) (Fig. 9a)). Their pores are also significantly bigger (Fig. 8.10a)).

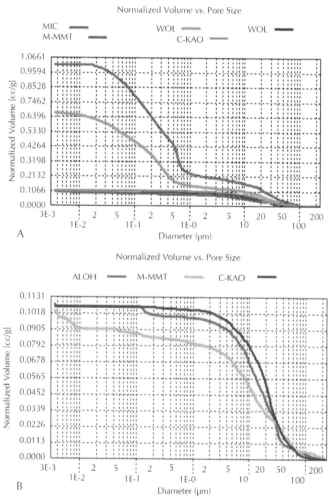

FIGURE 8.9   (a) The normalized pore volume vs. pore diameter for composites subjected to slow heating from room temperature to 1000 °C during 2 h; (b) Graph presents precisely differences in the characteristics for samples containing aluminum hydroxide (ALOH), surface treated montmorillonite (M-MMT) and calcined kaolin (C-KAO).

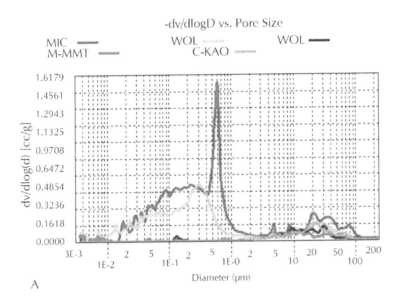

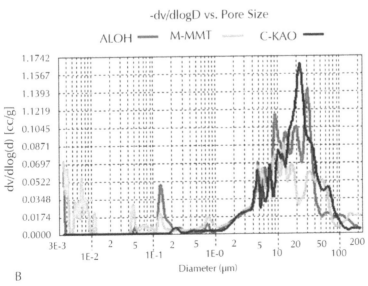

FIGURE 8.10 (a) The normalized differential curve dV/log(d) vs. pore diameter for composites slowly ceramized by heating from room temperature to 1000 °C during 2 h. (b) Graph presents precisely differences in the characteristics for samples containing aluminum hydroxide (ALOH), surface treated montmorillonite (M-MMT) and calcined kaolin (C-KAO).

8.4 SUMMARY AND CONCLUSIONS

Vulcanization kinetics of the composites studied are very similar, excluding sample filled with surface modified montmorillonite (M-MMT), for which torque values are lower, probably due to plasticization effect resulting from high concentration of organic modifier (quaternary ammonium salt). The M-MMT sample differs from other composites also in terms of the time of vulcanization. Both, scorch (t_{02}) and vulcanization (t_{90}) time of this sample are longer. It could be caused by acidic character of montmorillonite particles surface obtained as a result of Hoffmann beta-elimination of quaternary ammonium cation from the surface of particles [19, 25] during mixing of compounds. Acidic components affect the mechanism of peroxide crosslinking, leading to creation of non-active ions instead of peroxide radicals [23–25].

Elongation at break (EB), tensile strength (TS), stress at 100% (SE_{100}), 200% (SE_{200}) and 300% (SE_{300}) of elongation of composites before ceramization were generally similar except the M-MMT sample. In its case moduli at elongation and tensile strength values are lower, probably due to the plasticization effect originated from the presence of organic modifier. Tear strength of samples differs from 8.2 N/mm for C-KAO to 15.6 N/mm for M-MMT composite. It is likely that the presence of organofilizer influences positively TES of composites due to decreasing stiffness of vulcanizates, allowing dissipation of energy during tear deformation. Hardness of composites depends not only on the presence of organofilizer but also on the type of mineral filler added. However the difference is not significant.

Mechanical strength of composites after ceramization strongly depends on the type of refractory filler added. The best mechanical properties after isothermal ceramization at 600, 800 and 1000 °C present composite containing mica (MIC). After slow heating from room temperature to 1000 °C the strongest showed to be the composites filled with surface M-MMT, or MIC. Samples subjected to slow heating were generally shrinking except the MIC one, which diameter increased by 2 mm. This fact can adversely affect cable covers, which are likely to loose continuity on fire.

Porosimetry and SEM analysis demonstrate that type of refractory filler effects strongly morphology of the composite after ceramization. Only two samples, either containing mica (phlogopite) or wollastonite (WOL and MIC) are able to create nanoporous structure which can strongly improve thermal insulation of ceramic phase, what is very important from the point of view of remaining functioning of electric installations. A ceramized layer should be able to protect copper wire from outside heating during fire.

This study has shown that the best composition of fillers promoting ceramization of silicone rubber-based composites consists of boron oxide (fluxing agent) and mica (refractory filler). Good mechanical properties and processability, in combination with very good mechanical properties and nanoporous structure after ceramization give to this composite large industrial implementation capacity.

8.5 ACKNOWLEDGEMENTS

This work was supported by the EU Integrity Fund, project POIG 01.03.01-00-067/08-00.

KEYWORDS

- **Ceramization**
- **Composites**
- **Fillers**
- **Morphology**
- **Silicone rubber**

REFERENCES

1. Hamdani, S., Longuet, C., Lopez-Cuesta, J-M., & Ganachaud, F. (2010). Calcium and aluminum-based fillers as flame-retardant additives in silicone matrices. I. Blend preparation and thermal properties, *Polym Degr Stabil, 95,* 1911–1919.
2. Hamdani-Devarennes, S., Pommier, A., Longuet, C., Lopez-Cuesta, J-M., & Ganachaud, F. (2013). Calcium and aluminum-based fillers as flame-retardant additives in silicone matrices. II. Analyzes on composite residues from an industrial-based pyrolysis test, *Polym Degr Stabil 2011, 96,* 1562–1572.
3. Hamdani-Devarennes, S., Longuet, C., Sonnier, R., Ganachaud, F., & Lopez-Cuesta, J.-M. Calcium and aluminum-based fillers as flame-retardant additives in silicone matrices. III. Investigations on fire reaction, *Polym Degr Stabil, 98,* 2021–2032.

4. Mansouri, J., Wood, C. A., Roberts, K., Cheng, Y. B., & Burford, R. P. (2007). Investigation of the ceramifying process of modified silicone-silicate compositions. *J Mater Sci, 42,* 6046–6055.

5. Hanu, L. G., Simon, G. P., Mansouri, J., Burford, R. P., & Cheng, Y. B. (2004). Development of polymer-ceramic composites for improved fire resistance. *J Mater Process Tech, 153–154,* 401–407.

6. Bieliński, D. M., Anyszka, R., Pędzich, Z., & Dul, J. (2012). Ceramizable silicone rubber-based composites. Int *J Adv Mater Manuf Charact, 1,* 17–22.

7. Hanu, L. G., Simon, G. P., Cheng, Y. B. (2006). Thermal stability and flammability of silicone polymer composites. *Polym Degr Stabil, 91,* 1373–1379.

8. Pędzich, Z., Bukanska, A., Bieliński, D. M., Anyszka, R., Dul, J., & Parys, G. (2012). Microstructure evolution of silicone rubber-based composites during ceramization at different conditions. Int *J Adv Mater Manuf Charact, 1,* 29–35.

9. Pędzich, Z., & Bieliński, D. M. (2010). *Microstructure of silicone composites after ceramization Compos. 10,* 249–254.

10. Dul, J., Parys, G., Pędzich, Z., Bieliński, D. M., & Anyszka, R. (2012). Mechanical properties of silicone-based composites destined for wire covers. Int *J Adv Mater Manuf Charact, 1,* 23–28.

11. Hamdani, S., Longuet, C., Perrin, D., Lopez-Cuesta, J-M., & Ganachaud, F. (2009). Flame retardancy of silicone-based materials. *Polym Degr Stabil, 94,* 465–495.

12. Mansouri, J., Burford, R. P., Cheng, Y. B., & Hanu, L. (2005). Formation of strong ceramified ash from silicone-based composites. *J Mater Sci, 40,* 5741–5749.

13. Mansouri, J., Burford, R. P., & Cheng, Y. B. (2006). Pyrolysis behavior of silicone-based ceramifying composites. *Mater Sci Eng A, 425,* 7–14.

14. Hanu, L. G., Simon, G. P., & Cheng, Y. B. (2005). Preferential orientation of muscovite in ceramifiable silicone composites. *Mater Sci Eng A, 398,* 180–187.

15. Xiong, Y., Shen, Q., Chen, F., Luo, G., Yu, K., & Zhang, L. (2012). High strength retention and dimensional stability of silicone/alumina composite panel under fire. *Fire Mater, 36,* 254–263.

16. Pędzich, Z., Anyszka, R., Bieliński, D. M., Ziąbka, M., Lach, R., & Zarzecka-Napierała, M. (2013). Silicon-basing ceramizable composites containing long fibers. *J Mater Sci Chem Eng, 1,* 43–48.

17. Bieliński, D. M., Anyszka, R., Pędzich, Z., Parys, G., & Dul, J. (2012). Ceramizable silicone rubber composites. Influence of type of mineral on ceramization. *Compos, 12,* 256–261.

18. Pędzich, Z., Bukańska, A., Bieliński, D. M., Anyszka, R., Dul, J., & Parys, G. (2012). Microstructure evolution of silicone rubber-based composites during ceramization in different conditions. *Compos, 12,* 251–255.

19. Morgan, A. B., Chu, L. L., & Harris, J. D. (2005). A flammability performance comparison between synthetic and natural clays in polystyrene nanocomposites. *Fire Mater, 29,* 213–229.

20. Wang, T., Shao, H., & Zhang, Q. (2010). Ceramifying fire-resistant polyethylene composites. Adv *Compos Lett, 19,* 175–179.

21. Shanks, R. A., Al-Hassany, Z., & Genovese, A. (2010). Fire-retardant and fire-barrier poly(vinyl acetate) composites for sealant application. *Express Polym Lett, 4,* 79–93.

22. Thomson, K. W., Rodrigo, P. D. D., Preston, C. M., & Griffin, G. J. (2006). *Ceramifying polymers for advanced fire protection coatings.* Proceedings of European Coatings Conference, 2006.15th September, Berlin, Germany.
23. Heiner, J., Stenberg, B., & Persson, M. (2003). Crosslinking of siloxane elastomers. *Polym Test, 22,* 253–257.
24. Ogunniyi, S. D. (1999). Peroxide vulcanization of rubber. *Prog Rubber Plast Tech, 15,* 95–112.
25. Anyszka, R., Bieliński, D. M., & Kowalczyk, M. (2013). Influence of dispersed phase selection on ceramizable silicone composites cross-linking. *Elastom, 17,* 16–20.

CHAPTER 9

INVESTIGATION OF RUBBER WITH MICRODISPERSED WASTES OF SILICON CARBIDE

V. S. LIPHANOV, V. F. KABLOV, S. V. LAPIN, V. G. KOCHETKOV, O. M. NOVOPOLTSEVA, and G. E. ZAIKOV

CONTENTS

ABSTRACT

To create polymer materials exploiting in extreme conditions, it is required to use new components (including fillers) providing the flow of physical and chemical transformations that improve the operational stability of the materials. One of the problem solutions is the application of such promising fillers as high-dispersity silicon carbide in elastomer compositions. A low cost source of microdispersed silicon carbide can be slurries produced after grinding with the abrasive tool based on silicon carbide.

The paper considers the possibility of using microdispersed silicon carbide along with slurries as a functional filler in fire and heat resistant elastomer compositions. It has been shown that slurries formed after grinding and applied together with microdispersed silicon carbide can be used to effectively enhance the fire resistance of elastomer materials and make them cheaper.

9.1 INTRODUCTION

The Investigation of polymer materials for extreme operating conditions needs the new components that ensure the flow of physical and chemical transformations for increasing the operational resistance.

One of important components in elastomeric materials is the filler (carbon black, silica, etc.) The function of the filler is usually the improvement of mechanical properties (such as strength, hardness etc.). In extreme conditions, when temperatures are near and above the temperature of the material performance functionally active fillers can play a stabilizing role during the destruction of the material because of high temperatures [1, 2].

One effective solution is to use sloughing and fine fillers [5–8] and also highly dispersed silicon carbide [3–7].

Silicon carbide is one of the most promising materials. It has found application in many industries due to its high hardness and inertness to many corrosive environments. This material is now used to produce abrasive tools, as a filler for fire retardant materials, protecting covers of nuclear fuel, semiconductors and refractory composites. It is an advanced material for highly integrated devices microwave electronics, operating at high temperatures, high electric fields and high frequencies.

Silicon carbide (SiC) is a product of a chemical compound of carbon and silicon at high temperature [8]. It contains 70.04% of silicon and 29.96% carbon. The density is 3.1–3.2 g/cm^3; microhardness 3000–3300 kg/mm^2; hardness on the Mohs scale of more than 9.

Chemically pure silicon carbide is colorless, technical one can be found in a variety of colors from black to green and has a metallic luster.

The material has a lot of structural polytypes. The silicon carbide atoms are in state of sp3-hybridization and form a bond of a tetrahedron. In the crystal lattice of silicon carbide the short-range order is always the same but the long-range can differ that is why there are many polytypes of this material. The structural difference causes difference in physical and chemical properties (e.g., thermal resistance, electrical and optical characteristics). It makes one or another polytype being more preferred for different application.

Silicon carbide is a semiconductor material that is why it is a potential catalyst of thermal oxidation and pyrolysis processes. The silicon carbide particles have sharp corners and it allows to expect the appearance of physical and chemical activity in the processes of adsorption and chemical reactions (due to the presence of unpaired electrons and excess surface energy). Silicon carbide also can be used as so called microbarrier because of its plastic forms on the surface layers of the material. But the usage of silicone carbide in elastomeric materials is poorly understood.

A cheap source of microfine silicon carbide is a sludge, which is produced mainly while grinding with an abrasive tool based on silicon carbide. While grinding the silicon carbide particles are crushed to a smaller particles. Sludge also contains grinded metal microparticles and a small amount of surface-active substances.

9.2   THE PURPOSE OF THE RESEARCH

The purpose of the research is the investigation of the possibility to use the microfine silicon carbide – a basic material of a grinding sludge as a functionally active filler in the fire resistant elastomeric materials.

9.3 MATERIALS AND METHODS

The objects of the study are styrene-butadiene rubber vulcanizates SKMS-30ARKM 15 with sulfuric vulcanizing group [4–9]. Mixtures were prepared in laboratory rollers 160×320 mm. Vulcanization was carried out at a temperature of 145 °C.

The studied compositions are shown in Table 9.1.

TABLE 9.1   Test Compounds

Ingredient name	Weight parts/100 parts of rubber
Rubber SKMS-30ARKM-15	100
Carbon black P324	40

The research of the vulcanization kinetics of rubber compounds was carried out in accordance with GOST 12535–84 (Rubber compounds. The method for determining the cure characteristics with vulcametric) with Monsanto 100 rheometer. Physical and mechanical properties of the vulcanizates were determined by tensile testing machine IRI-60 according to GOST 270–75–75 (Rubber. The method of determining the elastic properties of tensile strength.). The hardness was estimated according to GOST 263–75 (Rubber. Determination of Shore A hardness.). The abrasion resistance was determined according to GOST 426–77 (Rubber. Method for the determination of abrasion resistance when sliding) on the Grasselli machine. Microscopic studies and determination of elemental composition were carried out on double-beam scanning electron microscope (Verse 3D).

9.4 RESULTS

Table 9.2 shows the rheological and vulcametrical parameters of the mixtures. As it follows from the table an optimal combination of rubber mixture is #3, it contains 20 weight parts of carbon black and 20 weight parts of grinding sludge. This combination of carbon black and silicon carbide increases the cure rate while increasing the induction period of vulcanization (time of vulcanization start). Increasing the amount of silicon carbide leads to faster curing that can be caused by the catalytic property of silicon carbide

TABLE 9.2 Vulcanization Properties of Rubber Mixtures*

Properties	1	2	3	A	Б
The minimum torque (ML), N·m	1.37	1.23	1.03	0.75	0.96
Maximum torque (MH), N·m	8.15	7.60	6.64	5.41	6.23
The difference in torque (ΔM), N·m	6.78	6.37	5.61	4.66	5.27
Start time of vulcanization (t$_s$), min	5.7	5.7	6.37	10.1	5.0
Time to reach 50% of cure (τ_{50}), min	13.0	11.6	12.0	14.6	8.7
The optimum cure time (τ_{90}), min	25.0	21.0	20.1	21.5	15.0
Vulcanization speed indicator (R$_v$), min-1	5.18	6.54	7.28	8.77	10.0

*The vulcanization temperature is 145 °C.

Physical and mechanical properties of vulcanizates are presented in Table 9.3. The results of flame resistance testing of the developed compositions are quiet interesting. The warm-up time of the backside of the sample to a temperature of 60°C was determined by the shims with 50 mm diameter and 6 mm thickness.

TABLE 9.3 Physical and Mechanical Properties of the Vulcanized Rubber

Parameter	1	2	3	A	Б
Vulcanization mode 145 °C × 25 min					
Apparent stress under 100% extension (f$_{100}$), MPa	2.06	2.20	1.44	0.97	0.97
Apparent stress under 300% extension, (f$_{300}$), MPa	9.4	8.8	4.3	1.56	1.40
Relative strength (fp), MPa	23.2	20.8	16.7	2.15	6.9
Elongation (ε), %	600	540	620	420	730
Permanent elongation (θ), %	20	12	13	4	20
Density, кг/м³	1110	1160	1170	1240	1600
Shore hardness, conventional unit	51	49	42	33	37
Attritive, α, m³/TDzh	73.3	63.2	68.3	94.6	160.3

The results of flame resistance testing of the developed compositions are quiet interesting. The warm-up time of the backside of the sample to a temperature of 60°C was determined by the shims with 50 mm diameter and 6 mm thickness.

When the flame burns the sample with silicone carbone the tight fire resistant chark appears, it protects the sample from burning itself. The plastic form of silicone carbide particles generates the barrier layer that also protects sample from the flame.

The plastics of silicone carbide can be seen on the surface of chark. As silicone carbide is a heat-resistant and a very hard to oxidize, so the barrier layer effectively protects the sample from burning through.

To estimate the heat resistance of the developed vulcanizates when heating with plasma torch the temperature of the non-heating surface was measured. After the silicone carbide was impregnated the time the sample reaches the temperature of 60°C increased from 33 to 60 min.

9.5 CONCLUSIONS

The researches are showing that the grinding sludge with microfine silicon carbide can be used to increase the efficiency of fire resistant elastomeric materials while decreasing their cost.

KEYWORDS

- **Elastomers**
- **Fillers**
- **Fire resistance**
- **Silicon carbide**

REFERENCES

1. Hamdani, S., Longuet, C., Lopez-Cuesta, J-M., & Ganachaud, F. (2010). Calcium and aluminum-based fillers as flame-retardant additives in silicone matrices. I. Blend preparation and thermal properties. *Polym Degr Stabil, 95,* 1911–1919.

2. Hamdani-Devarennes, S., Pommier, A., Longuet, C., Lopez-Cuesta, J-M., & Ganachaud, F. (2011). Calcium and aluminum-based fillers as flame-retardant additives in silicone matrices. II. Analyzes on composite residues from an industrial-based pyrolysis test. *Polym Degr Stabil, 96,* 1562–1572.

3. Hamdani-Devarennes, S., Longuet, C., Sonnier, R., Ganachaud, F., & Lopez-Cuesta, J-M. (2013). Calcium and aluminum-based fillers as flame-retardant additives in silicone matrices. III. Investigations on fire reaction. *Polym Degr Stabil, 98,* 2021–2032.

4. Mansouri, J., Wood, C. A., Roberts, K., Cheng, Y. B., & Burford, R. P. (2007). Investigation of the ceramifying process of modified silicone-silicate compositions. *J Mater Sci, 42,* 6046–6055.

5. Hanu, L. G., Simon, G. P., Mansouri, J., Burford, R. P., & Cheng, Y. B. (2004). Development of polymer-ceramic composites for improved fire resistance. *J Mater Process Tech, 153–154,* 401–407.

6. Bieliński, D. M., Anyszka, R., Pędzich, Z., & Dul, J. (2012). Ceramizable silicone rubber-based composites. *Int J Adv Mater Manuf Charact, 1,* 17–22.

7. Hanu, L. G., Simon, G. P., & Cheng, Y. B. (2006). Thermal stability and flammability of silicone polymer composites. *Polym Degr Stabil, 91,* 1373–1379.

8. Pędzich, Z., Bukanska, A., Bieliński, D. M., Anyszka, R., Dul, J., & Parys, G. (2012). Microstructure evolution of silicone rubber-based composites during ceramization at different conditions. *Int J Adv Mater Manuf Charact, 1,* 29–35.

9. Pędzich, Z., & Bieliński, D. M. (2010). Microstructure of silicone composites after ceramization Compos, 10, 249–254.

10. Dul, J., Parys, G., Pędzich, Z., Bieliński, D. M., & Anyszka, R. (2012). Mechanical properties of silicone-based composites destined for wire covers. *Int J Adv Mater Manuf Charact, 1,* 23–28.

11. Hamdani, S., Longuet, C., Perrin, D., Lopez-Cuesta, J-M., & Ganachaud, F. (2009). Flame retardancy of silicone-based materials. *Polym Degr Stabil, 94,* 465–495.

12. Mansouri, J., Burford, R. P., Cheng, Y. B., & Hanu, L. (2005). Formation of strong ceramified ash from silicone-based composites. *J Mater Sci, 40,* 5741–5749.

13. Mansouri, J., Burford, R. P., & Cheng, Y. B. (2006). Pyrolysis behavior of silicone-based ceramifying composites. *Mater Sci Eng A, 425,* 7–14.

14. Hanu, L. G., Simon, G. P., & Cheng, Y. B. (2005). Preferential orientation of muscovite in ceramifiable silicone composites. *Mater Sci Eng A, 398,* 180–187.

15. Xiong, Y., Shen, Q., Chen, F., Luo, G., Yu, K., & Zhang, L. (2012). High strength retention and dimensional stability of silicone/alumina composite panel under fire. *Fire Mater, 36,* 254–263.

16. Pędzich, Z., Anyszka, R., Bieliński, D. M., Ziąbka, M., Lach, R., & Zarzecka-Napierała, M. (2013). Silicon-basing ceramizable composites containing long fibers. *J Mater Sci Chem Eng, 1,* 43–48.

17. Bieliński, D. M., Anyszka, R., Pędzich, Z., Parys, G., & Dul, J. (2012). Ceramizable silicone rubber composites. Influence of type of mineral on ceramization. *Compos, 12,* 256–261.

18. Pędzich, Z., Bukańska, A., Bieliński, D. M., Anyszka, R., Dul, J., & Parys, G. (2012). Microstructure evolution of silicone rubber-based composites during ceramization in different conditions. *Compos, 12,* 251–255.

19. Morgan, A. B., Chu, L. L., & Harris, J. D. (2005). A flammability performance comparison between synthetic and natural clays in polystyrene nanocomposites. *Fire Mater, 29,* 213–229.

20. Wang, T., Shao, H., & Zhang, Q. (2010). Ceramifying fire-resistant polyethylene composites. *Adv Compos Lett, 19,* 175–179.

21. Shanks, R. A., Al-Hassany, Z., & Genovese, A. (2010). Fire-retardant and fire-barrier poly(vinyl acetate) composites for sealant application. *Express Polym Lett, 4,* 79–93.

22. Thomson, K. W., Rodrigo, P. D. D., Preston, C.M., & Griffin. G. J. (2006). *Ceramifying polymers for advanced fire protection coatings.* Proceedings of European Coatings Conference 2006, 15th September, Berlin, Germany.

23. Heiner, J., Stenberg, B., & Persson, M. (2003). Crosslinking of siloxane elastomers. *Polym Test, 22,* 253–257.

24. Ogunniyi, S. D. (1999). Peroxide vulcanization of rubber. *Prog Rubber Plast Tech, 15,* 95–112.

25. Anyszka, R., Bieliński, D. M., & Kowalczyk, M. (2013). Influence of dispersed phase selection on ceramizable silicone composites cross-linking. *Elastom, 17,* 16–20

26. Grishin, B. S. (2010). Materialy rezinovoy promyshlennosti (informatsionno-analiticheskaya baza dannykh). Ch. *2* [Rubber Materials Industry (information-analytical database)]. *Kazan.* 488 p.

27. Kornev, A. E. (2009). *Tekhnologiya elastomernykh materialov* [Elastomeric materials technology] Moscow. Istek Publ. 504 p

28. Kablov, V. F., & Novopoltseva, O. M. (2013). Vliyaniye napolnitelya perlit na teplostoykost rezin na osnove etilenpropilendiyenovogo kauchuka [Effect of the heat-resistant filler perlite on ethylene-propylene rubbers] *Modern Problems of Science and Education, (3),* 444.

29. Kablov, V. F., & Novopoltseva, O. M. (2009). Materialy i sozdaniye retseptur rezinovykh smesey dlya shinnoy i rezinotekhnicheskoy promyshlennosti. [Materials and creating recipes rubber compounds for tire and rubber industry] *Tekhnologiya pererabotki plasticheskikh mass i elastomerov* (Processing technology of plastics and elastomers) *Volgograd,* 321.

30. Kablov, V. F., & Novopoltseva, O. M. (2013). Razrabotka i issledovaniye ogneteplozashchitnykh materialov s vspuchivayushchimisya i mikrovoloknistymi napolnitelyami s elementorganicheskimi modifikatorami dlya ekstremalnykh usloviy ekspluatatsii [Development and research of fire retardant intumescent materials and micro-fiber fillers with organoelement modifiers for extreme conditions] Tezisy dokladov III-ey Vserossiyskoy konferentsii Kauchuk i rezina – (2013)*: traditsii i novatsii (Abstracts of the III-rd Russian Conference "Caoutchouc and rubber"),* Moscow, 28–30.

31. Kablov, V. F., & Novopoltseva, O. M. (2012). *Teplozashchitnyye pokrytiya, soderzhashchiye perlit [Thermal barrier coatings containing perlite] Mezhdunarodnyy zhurnal prikladnykh i fundamentalnykh issledovaniy (International Journal of Applied and Basic Research), (1),* 174–175.

32. Novakov, I. A., & Kablov, V. F. (2011). Vliyaniye napolniteley, modifitsirovannykh metallami peremennoy valentnosti, na vysokotemperaturnoye stareniye rezin na osnove etilen-propilenovogo kauchuka [Effect of fillers modified with transition metals at high temperature aging of rubbers based on ethylene-propylene rubber] *Izvestiya*

Volgogradskogo gosudarstvennogo tekhnicheskogo universiteta (News of the Volgograd State Technical University), Volgograd, *2(8)*, 102–105.

33. Knunyantsa, I. L. (1990). *Khimicheskaya entsiklopediya.* T. 2 [Chemical encyclopedia V.2], Moscow, Sovetskaya entsiklopediya Publ, 673.

34. Reznichenko, S. V., & Morozova, Yu. L. (2012). *Bolshoy spravochnik rezinshchika.* Ch.1. Kauchuki i ingrediyenty [Great reference book of rubber maker. P.1. Rubbers and ingredients], Moscow, Tekhinform Publ, 744 p.

PART 3

MODIFICATION OF ELASTOMERS AND THEIR COMPONENTS

CHAPTER 10

EFFECT OF THERMO-MECHANOCHEMICAL CHANGES OF NATURAL RUBBER ON SOME CHARACTERISTIC OF RUBBER COMPOUNDS AND VULCANIZED RUBBERS

I. A. MIKHAYLOV, YU. O. ANDRIASYAN, R. JOZWIK, G. E. ZAIKOV, and A. A. POPOV

CONTENTS

ABSTRACT

Thermo-mechanochemical changes of natural rubber SVR 3L under treatment internal mixer at self-heating have been studied. Effect of molecular mass and content of gel-fraction of natural rubber is shown. Properties of rubber compounds and vulcanized rubber are presented in this chapter.

10.1 AIMS AND BACKGROUND

Now the world elastomeric market consists of the natural rubber (NR) – 40% and synthetic rubbers – 60%. According to forecasts of experts the tendency of increase in a share of NR is observed. It is supposed that by 2015–2020 its share will make 50%. For the purpose of expansion of a scope of natural rubber manufacturing of NR with the content of chlorine from 0.5 to 15% is of interest since it is known that such content of halogen doesn't worsen flexural properties of rubber.

Halide modification of polymers and natural rubber in particular together with obtaining of halogen-containing polymers with a help of synthesis is one of intensively developing direction in the field of obtaining chlorine-containing polymers. In result of carrying out halide modification of polymers, which have technologically smoothly, large capacity industrial production, elastomer materials and composites are managed to obtain with wide complex of a new specific properties: high adhesion, fire-, oil-, gasoline-, heat resistance, ozone resistance, incombustibility, resistance to influence of corrosive environments and microorganisms, high strength, gas permeability, etc.

10.2 RESULTS AND DISCUSSION

In pervious works on haloid mechanochemical modification of synthetic analog of NR, isoprene SKI-3 [1–3] rubber, it was established that the size of the molecular weight (M_η) and the contents fraction gel (C_g, %) have essential impact on depth of course of reaction halogen accession to thermomechanical activated macromolecules of rubber.

The purpose of this work is studying of features of mechanochemical transformations of NR in the course of its processing in a two-rotor high-speed rubber-mixer for identification of the most optimum areas of carrying out haloid mechanochemical modification.

At the first stage we determined the size of M_η and the contents gel fraction of samples of NR of main suppliers of this natural polymer to the world market.

Certain structural parameters of NR and the producer country are specified in Table 10.1.

TABLE 10.1   Structural Parameters of NR

Type of rubber	$[\eta]$	$M_\eta \times 10^{-4}$	C_g, %
SVR 3L Vietnam	6.27	138	19.5
NKHC Cameroon	5.9	127.5	13.3
SMR GP Malaysia	6.4	144	2.5

From the data provided in the table it is visible that rubber of the Vietnamese production (SVR 3L) [4] most fully meets earlier designated requirements for M_η and the contents of gel fraction.

Further the process of conversion of natural rubber of the SVR 3L brand (Vietnam) was studied and structural parameters of samples of the processed natural rubber and property of elastomeric compositions on their basis are determined.

Mechanical processing of NR was carried out on a laboratory two-rotor rubber-mixer of type RVSD-01-60 (with friction 1:1.5). Duration of machining of SVR 3L rubber samples was 5, 10, 20, 30, 40, 50 and 60 min. Skilled samples of rubber were overworked in a self-heating mode. There were studied M_η change, the contents gel fraction and temperatures of rubbers depending on duration of mechanical influence.

Results of research of thermomechanochemical transformations of SVR 3L rubber are presented on Fig. 10.1.

From the provided data it is visible that at thermomechanical processing of NR mechanical degradation proceed generally at an initial stage of processing (up to 10 min) in the field of rather low temperatures (from 20 to 110 °C). In this time interval change of molecular weight from 140 to 68×10^4 and contents gel fraction from 20 to 5% is observed.

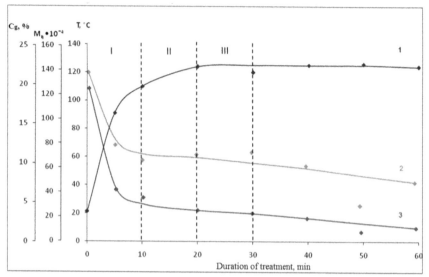

FIGURE 10.1   Temperature (1), average molecular weight (2) and the contents fraction gel (3) depending on duration of treatment of mechanical processing of SVR 3L (Vietnam).

At processing times from 10 to 60 min little change of M_η from 68 to 50×10^4, and gel fraction from 5 to 3% is observed. Temperature of processed rubber in the range from 10 till 60 min of processing changes from 110 to 130 °C.

By consideration of the results received at processing of NR from 10 till 60 min, it is possible to assume that thermomechanoactivation processes without a rupture of macromolecules and the partial (small) thermally activated destruction of macromolecules of rubber and its gel fraction generally prevail here.

For the purpose of studying of influence of structural parameters of NR samples processed in various temperature time intervals on properties of rubber compounds and rubbers on their basis, within the standard filled compounding for NR [5] rubber mixes were made and their vulcanizing properties and physicomechanical properties of vulcanizers were defined. Results of the conducted researches are given in Table 10.2.

TABLE 10.2 Influence of Duration of Treatment SVR 3L on Rubber Compounds Vulcanizing Properties and Physicomechanical Properties of Vulcanizers

Index	Duration of treatment SVR 3L, min							
	0	5	10	20	30	40	50	60
Rotational moment, dN*m								
M_{min}	6.7	6.5	6	6	3.5	2.8	2.5	2.5
M_{max}	29.7	28	27.5	27	24.5	22.2	22.5	21.5
Initial time of vulcanization, min	2	1.5	2	2.2	2.5	1.5	2.5	2.5
Optimum vulcanization time, min	7	7	7	8	8	7	8	8
Vulcanization rate, %/min	20	18.2	20	17.2	18.2	18.2	18.2	18.2
Conventional tensile strength 200%, MPa	1.6	1.5	1.1	1.3	1.2	1.1	1.1	1.1
Conventional tensile strength 500%, MPa	17.3	16.5	15.4	16.6	14.4	13.6	13.6	12.8
Conventional tensile strength, MPa	20.6	20.5	20.2	19.3	16.3	15	14.2	13.3
Conventional breaking elongation, %	675	660	620	610	600	600	575	570

From the data provided in Table 10.2 it is visible that with increase of processing time plasticity of rubber mixes increases caused by reduction the $M_{outgoing}$ and M_{min}, connected with decrease of rubber molecular weight to course of degradation processes. Essential influence on such vulcanizing properties as time of the beginning of curing, optimum time of curing and speed of curing isn't observed. Some fall of value of the maximum torque (M_{max}) with increase in overtime of NR processing is noticeable that points to reduction of strength properties of vulcanizers of rubber samples.

By consideration of physicomechanical properties at increase in overtime of rubber falling of strength characteristics (conditional tension is revealed at 200%, 500% and conditional durability at stretching), insignificant decrease in relative lengthening and increase in relative residual lengthening after a gap are observed. It should be noted that till 30 min of processing inclusive, change of the parameters specified above is insignificant while at overtime, 40, 50 and 60 min their sharp change is observed.

On the basis of the critical data analysis obtained as a result of made by us experiment three areas of mechanochemical haloid modification of natural SVR 3L rubber are allocated:

Area I– rubber overtime from 0 to 10 min (mechanodestruction of macromolecules and rubber gel fraction);

Area II – rubber overtime from 10 to 20 min (mechanoactivation and thermally activated mechanodestruction);

Area III – rubber overtime from 20 to 30 min (prevailing mechanoactivation and somewhat thermally activated mechanodestruction).

The most probable processes proceeding in the above-stated temperature time intervals, are presented on the following schemes:

I: Mechanodestruction process

$$\underset{|}{\overset{CH_3}{}} \quad \text{mech. treatment} \quad \underset{|}{\overset{CH_3}{}}$$
$$\sim CH_2\text{-}C\text{=}CH\text{-}CH_2\sim \longrightarrow \sim CH_2\text{-}C\text{=}\overset{.}{C}H + \overset{.}{C}H_2\sim$$

II: Mechanodestruction and mechanoactivation (change of valent corners of the C-C communication without its gap)

$$\underset{|}{\overset{CH_3}{}} \quad \text{mech. treatment} \quad \underset{|}{\overset{CH_3}{}}$$
$$\sim CH_2\text{-}C\text{=}CH\text{-}CH_2\sim \longrightarrow \sim CH_2\text{-}C\text{=}\overset{.}{C}H + \overset{.}{C}H_2\sim \cdot \text{mechanodestruction}$$

$$\underset{|}{\overset{CH_3}{}} \quad \text{mech. treatment} \quad \underset{|}{\overset{CH_3}{}}$$
$$\sim CH_2\text{-}C\text{=}CH\text{-}CH_2\sim \longrightarrow \sim CH_2\text{-}C\text{=}CH\text{----}CH_2\sim \cdot \text{mechanoactivation}$$

III: Mechanoactivation

$$\underset{|}{\overset{CH_3}{}} \quad \text{mech. treatment} \quad \underset{|}{\overset{CH_3}{}}$$
$$\sim CH_2\text{-}C\text{=}CH\text{-}CH_2\sim \longrightarrow \sim CH_2\text{-}C\text{=}CH\text{----}CH_2\sim$$

10.3 CONCLUSIONS

Thus the conducted researches allowed to establish nature of change of structural parameters of natural SVR 3L rubber (M_η and the contents gel fraction) depending on duration of mechanical processing of polymer, to study the influence of these parameters on properties of elastomeric compositions on the basis of rubbers subjected thermo-mechanical influence. Taking into account the nature of the proceeding mechanochemical processes observed at processing of natural rubber, and also properties of elastomeric compositions on the basis of these rubbers the most acceptable temperature time intervals of carrying out mechanochemical haloid modification are defined. It is supposed that processing of rubbers in the above-stated temperature time intervals in the presence of the chlorine-containing modifier has to reveal influence of mechanical degradation and mechanoactivation processes on depth of reaction of halogenation.

KEYWORDS

- **Cauotchouc**
- **Elastomer**
- **Mechanical chemistry**
- **Rubber**
- **Rubber compound**
- **Technology**

REFERENCES

1. Andriasyan, Yu O., Kornev, A. E., & Gyulbekyan, A. L. (2001). Theses of the conference report IX RAS *Degradation and Stabilization of Polymers* (Moscow 2001), 11.
2. Andriasyan, Yu O., Popov, A. A., Gyulbekyan, A. L., & Kornev, A. E. (2002). *Caoutchouc and Rubber*, 3, 4–6.
3. Andriasyan, Yu O., Popov, A. A., Gyulbekyan, A. L., & Kornev, A. E. (2002). *Caoutchouc and Rubber*, 4, 18–20.
4. Dumnov, S. E., Mikhailov, I. A., Andriasyan, Yu. O., Popov, A. A., Kashiricheva, I. I., & Kornev, A. E. (2007). Theses of reports. Seventh Annual International Conference of *Biochemical Physics* RAS – Colleges (Moscow 2007), 102–104.
5. Manual of Rubberier. *M.: Chemistry*. 1971, 608.

CHAPTER 11

RADIATION CROSSLINKING OF ACRYLONITRILE-BUTADIENE RUBBER—THE INFLUENCE OF SULFUR AND DIBENZOTHIAZOLE DISULFIDE ON THE PROCESS

KATARZYNA BANDZIERZ, DARIUSZ M. BIELINSKI,
ADRIAN KORYCKI, and GRAZYNA PRZYBYTNIAK

CONTENTS

ABSTRACT

Radiation crosslinking of elastomers has been receiving increasing attention. The reactions induced by high-energy ionizing radiation are very complicated and the mechanisms still remain not entirely comprehended. Ionizing radiation crosslinking of acrylonitrile-butadiene rubber, filled with 40 phr of silica, with incorporated sulfur crosslinking system was the object of study. To investigate the influence of components such as sulfur and crosslinking accelerator – dibenzothiazole disulfide (DM) on the process, a set of rubber samples with various sulfur to crosslinking accelerator ratio was prepared and irradiated with 50, 122 and 198 kGy. Crosslink density and crosslink structure were analyzed and mechanical properties of the rubber samples were determined. Inhibiting effect of DM and sulfur on the radiation crosslinking process was found. The rubber vulcanizates having sulfur in composition characterized themselves with hybrid crosslinks – both carbon–carbon and sulfide, with varying degree of sulfidity, depending on the sulfur and DM content. The presence of sulfide crosslinks increased tensile strength of the rubber samples.

11.1   INTRODUCTION

Radiation modification of polymer materials has been gaining increasing popularity, not only in academic research, but also in industrial applications [1]. Among numerous advantages of radiation modification method, the noteworthy issue is the simplicity to control the ionizing radiation dose, which is absorbed by the modified material, dose rate and energy of ionizing radiation. The resulting properties can be therefore 'tailored' and the whole process is highly controllable and repeatable.

Radiation crosslinking is an interesting alternative for thermal crosslinking [2–4] or its complement [5–7]. One of extensively studied polymers in respect to its radiation crosslinking, is acrylonitrile-butadiene rubber (NBR) [8–12], which belongs to group of polymers, which effectively crosslink on irradiation with ionizing radiation.

As a result of high-energetic irradiation, radicals are generated directly on polymer chains. By recombination, they form carbon–carbon (C-C) crosslinks between the chains. Due to the fact that radiation crosslinking leads to formation of C-C crosslinks and the mechanism is radical, it is often compared to peroxide crosslinking [3, 13]. It is noteworthy to enhance that the processes induced by ionizing radiation are very complicated and therefore not thoroughly understood [14]. C-C crosslinks provide good elastic properties and are resistant to thermal aging, but they are short, stiff and are do not provide satisfactory properties for dynamic loadings.

According to Dogadkin's theory, crosslinks of various structures provide better mechanical properties, owing to different lengths of the bridges between polymer chains, which do not break at the same time [15]. To provide optimal properties required for the end-use product, the researchers endeavor to design materials with hybrid (mixed) crosslinks. To obtain hybrid type network with both C-C and longer sulfide crosslinks, studies on thermal simultaneous crosslinking with two types of curatives, such as organic peroxide and sulfur crosslinking system, were carried out. The results generally showed lowered efficiency of crosslinking due to competing reactions, in which sulfur and crosslinking accelerator are involved in reactions with peroxide radicals [16–19].

In our previous research [20] concerning radiation crosslinking of NBR rubber with sulfur crosslinking system in composition, hybrid crosslink structure was proved to form upon irradiation. Inhibiting effect of sulfur crosslinking system on total crosslink density, formed in the irradiation process, was observed. To investigate in detail the contribution coming from particular components of the crosslinking system, such as rhombic sulfur and crosslinking accelerator DM, a set of samples with various ratios of sulfur and accelerator was prepared. The effect of these two components on radiation crosslinking process was studied by determination of basic mechanical properties, total crosslink density and analysis of crosslink structure.

11.2 EXPERIMENTAL PART

11.2.1 MATERIALS

Acrylonitrile – butadiene rubber Europrene N3325 (bound ACN content 33%) was supplied by Polimeri Europa (Italy). Precipitated silica Ultrasil VN3 was obtained from Evonik Industries (Germany). Vinyltrimethoxysilane U-611 was obtained from Unisil (Poland). Rhombic sulfur was provided by Siarkopol Tarnobrzeg (Poland) and zinc oxide, stearic acid and dibenzothiazole disulfide (DM), by Lanxess (Germany).

11.2.2   SAMPLES PREPARATION

Rubber mixes were prepared in two-stage procedure. In the first stage, rubber premixes of NBR, filled with 40 phr of precipitated silica and vinyltrimethoxysilane in amount of 10 wt. % of silica in the composite mix, were prepared with the use of Brabender Plasticorder internal micromixer (Germany) at temperature of mixing chamber of 120 °C, with rotors speed of 20 RPM during components incorporation and 60 RPM during 25 min lasting homogenization process. In the second stage, components of crosslinking system, such as zinc oxide, stearic acid, rhombic sulfur and dibenzothiazole disulfide (DM), were incorporated into the premix with David Bridge two-roll open mixing mill (UK) at 40°C and homogenized for 10 min. The samples composition is given in Table 11.1. A 1-mm rubber mixes sheets were compression molded in an electrically heated press at temperature of 110°C under pressure of 150 bar for 4 min.

TABLE 11.1   Composition of Rubber Mixes*

(Rubber mixes x/y)/(Component, wt. %)	0 / 0	0 / 1.5	2 / 0	2 / 1.5
NBR, Europrene N3325	100	100	100	100
Silica, Ultrasil VN3	40	40	40	40
Silane, U611	4	4	4	4
Zinc oxide, ZnO	5	5	5	5
Stearic acid	1	1	1	1
Rhombic sulfur, S_8	0	0	2	2
Dibenzothiazole disulfide, DM	0	1.5	0	1.5

*The samples are designated as x/y, where, x indicates the amount of sulfur and y indicates the amount of dibenzothiazole disulfide, respectively.

11.2.3 SAMPLES IRRADIATION

The molded rubber sheets were subjected to electron beam (EB) irradiation at Elektronika 10/10 linear electron accelerator (Russia), located at the Institute of Nuclear Chemistry and Technology (Poland). The absorbed doses were 50, 122 and 198 kGy. Irradiation process was carried out in air atmosphere at room temperature. The rubber sheets were placed horizontally in the front of pulsed, scanned beam. The total doses were obtained by multi-pass exposure (approx. 25 kGy per pass).

11.2.4 SAMPLES CHARACTERIZATION

11.2.4.1 CROSSLINK DENSITY DETERMINATION

Total crosslink density of the irradiated samples was determined taking advantage of equilibrium swelling in toluene and calculated on the basis of Flory-Rehner equation [21]. The Flory-Huggins interaction parameter used in the calculations for toluene – NBR rubber was 0.435 [22].

11.2.4.2. CROSSLINK STRUCTURE DETERMINATION

The crosslink structure was analyzed and quantified by thiol–amine analysis, which is based on treatment of the crosslinked material with a set of thiol–amine chemical probes, specifically cleaving particular crosslinks types [23]. Polysulfide crosslinks are cleaved by treatment of crosslinked rubber samples with 2-propanethiol (0.4 M) and piperidine (0.4 M) in toluene for 2 h under inert gas atmosphere (argon) at room temperature, while polysulfide and disulfide crosslinks can be cleaved by treatment under the same conditions with 1-dodecanethiol (1 M) in piperidine for 72 h.

11.2.4.3 MECHANICAL PROPERTIES

Mechanical tests were carried out with the use of 'Zwick 1435' universal mechanical testing machine (Germany), according to ISO 37. The crosshead speed was 500 mm/min and the temperature was 23±2 °C. Five dumbbell specimens were tested for each sample and the average is reported here.

11.3　RESULTS AND DISCUSSION

11.3.1　CROSSLINK DENSITY

The crosslink densities of samples irradiated with doses of 50, 122 and 198 kGy, calculated from equilibrium swelling in toluene, are presented in Fig. 11.1. For all samples studied, crosslink densities formed during EB irradiation process are increasing linear function of dose.

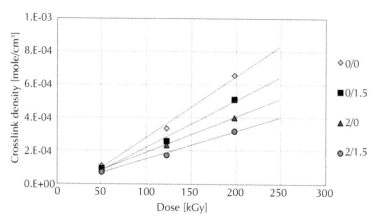

FIGURE 11.1　Total crosslink density as a function of ionizing radiation dose.

The inhibiting effect of DM and sulfur on the radiation crosslinking process was observed. According to experimental work, the inhibiting effect of sulfur (sample 2/0) and DM (sample 0/1.5) is not additive, comparing to corresponding inhibition coming from the same amount of sulfur and DM combined in one sample (2/1.5). The "experimental inhibition" (sample 2/1.5) is lower that the "theoretical inhibition" (summed up inhibition of samples 2/0 and 0/1.5), as shown in Fig. 11.2. The probable explanation of the fact can be sulfur – accelerator complex formed during sheets molding process at 110°C. The complex formed facilitate formation of sulfide crosslinks and possibly makes the reactions more effective –

sulfur is used rather for formation of bridges between the polymer chains, than for formation of cyclic structures modifying the chains.

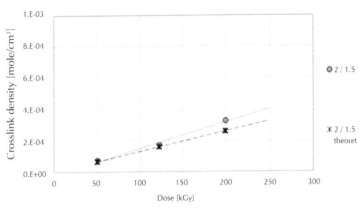

FIGURE 11.2 "Experimental inhibition" (solid line) and "theoretical inhibition" (dashed line) of radiation crosslinking process.

Inhibiting effect of DM arises from the presence of aromatic rings in its structure. The aromatic compounds are known to influence the radiation-induced modification by effect of resonance energy dissipation [24–25]. In the structural formula of DM, heteroatoms, such as sulfur and nitrogen are also present. The sulfur moieties are known to inhibit the effect of ionizing radiation action on matter [26–27], due to the fact that sulfur groups act as sinks of the radiation energy [28].

Rhombic sulfur itself also causes large inhibiting effect of subjected to ionizing radiation polymer. It has to be enhanced, that sulfur undoubtlessly is strong radiation-protecting agent, but probably the crosslinking efficiency is reduced because of intramolecular reactions, which result in modification of polymer chains by sulfur cyclic structures. The observed inhibiting effect of sulfur has therefore twofold contribution.

11.3.2 CROSSLINK STRUCTURE

For samples with rhombic sulfur in composition, the crosslink structure investigation was carried out (Fig. 11.3). Presence of both C-C and polysulfide crosslinks was proved. The C-C crosslinks are "regular" effect of polymer irradiation. The sulfide crosslinks were formed as a result of breakage of S-S bonds in highly puckered ring structure of rhombic sulfur by the action of accelerated electrons. The S-S bond energy is low – approx. 240 kJ/mole, what makes it susceptible to break and generate sulfur radicals [6], what consequently leads to sulfide crosslinks formation.

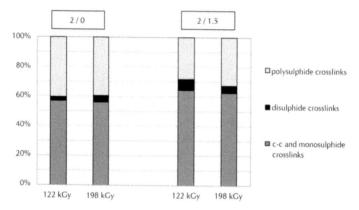

FIGURE 11.3　Crosslink structure of samples 2/0 and 2/1.5, irradiated with 122 and 198 kGy. Network density formed in the samples irradiated with a dose of 50 kGy was very low and the results on crosslink structure obtained from the thiol – amine analysis was not reliable.

The crosslink structure study showed that during irradiation of the sample containing sulfur, but without crosslinking accelerator (sample 2/0), the participation of polysulfide crosslinks in the total crosslink density is approx. 40%. The difference between the number of polysulfide crosslinks formed upon irradiation with 122 and 198 kGy is very little. In sample 2/1.5 in which both sulfur and crosslinking accelerator are present, the number of polysulfide crosslinks is lower than in sample 2/0, and it slightly increases with irradiation dose (from 28% for 122 kGy up to 32% for 198 kGy). The presence of complex of crosslinking accelerator with sulfur promoted thereby formation of shorter crosslinks.

11.3.3. MECHANICAL PROPERTIES

The mechanical properties of all samples studied are presented in Table 11.2.

TABLE 11.2 Mechanical Properties (SE100, SE200, SE300, TS, Eb) of Samples Irradiated with 50, 122 and 198 kGy

Sample	Dose [kGy]	Crosslink density [mol/cm³]	Mechanical properties				
			SE_{100} [MPa]	SE_{200} [MPa]	SE_{300} [MPa]	TS [MPa]	E_b [MPa]
0 / 0	50	$1.1×10^{-4}$	2.7	4.0	5.6	10.6	599
	122	$3.4×10^{-4}$	5.1	10.2	17.2	25.1	397
	198	$6.6×10^{-4}$	9.8	21.1	–	23.6	219
0 / 1.5	50	$9.3×10^{-5}$	2.0	2.9	4.0	7.6	670
	122	$2.6×10^{-4}$	3.9	7.6	12.5	20.8	441
	198	$5.1×10^{-4}$	7.5	15.8	–	23.7	277
2 / 0	50	$8.3×10^{-5}$	1.9	2.6	3.5	7.6	756
	122	$2.4×10^{-4}$	4.2	7.8	12.5	25.2	503
	198	$4.0×10^{-4}$	6.1	12.6	21.6	30.9	388
2 / 1.5	50	$7.3×10^{-5}$	2.1	2.7	3.5	6.5	756
	122	$1.7×10^{-4}$	3.0	5.3	8.2	17.4	549
	198	$3.2×10^{-4}$	4.8	9.7	16.3	26.7	429

In sample 2/0, the generated sulfur radicals inserted into polymer chains, forming long, polysulfide crosslinks, which have significant participation in the total crosslink density. The presence of polysulfide crosslinks is evident in mechanical properties – high tensile strength is provided by these long, labile bridges, which effectively dissipate the energy. Due to this effect, sample 2/0 showed the highest tensile strength among all analyzed samples (Fig. 11.4). The lowest value of tensile strength exhibited the sample 0/1.5. The presence of the DM not only inhibited the formation of crosslinks, but also considerably deteriorated the resulting mechanical properties of the rubber sample. Modification of the polymer chain by the products of DM transformation upon irradiation is probable.

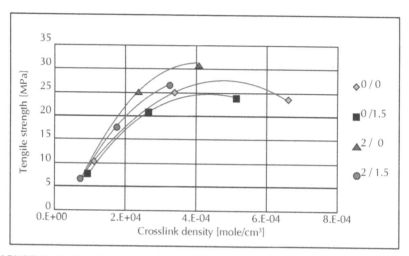

FIGURE 11.4   Tensile strength of samples as a function of crosslink density.

Tensile strength curve of sample containing both sulfur and DM, is located between the corresponding curves of samples containing solely DM or sulfur. Its tensile strength is higher than of sample without sulfur or DM (0/0), due to presence of mixed, diversified crosslinks in sample 2/1.5, and exclusively uniform C-C crosslinks in sample 0/0.

11.4 CONCLUSIONS

In our study, the influence of particular components of sulfur crosslinking system, such as rhombic sulfur and crosslinking accelerator DM, on the process of radiation crosslinking of NBR was investigated.

- Inhibition of radiation crosslinking by both sulfur and DM was proved. Due to complex nature of the investigated system and complicated processes induced by high-energy radiation, it is difficult to unambiguously identify the mechanisms responsible for the inhibiting effect.
- Irradiation of samples with sulfur in composition leads to formation of hybrid network type, characterizing itself with both short C-C crosslinks and longer sulfide ones. The presence of diversified crosslinks guarantee high tensile strength of the rubber samples.

The reactions induced in polymer matrix with sulfur crosslinking system in composition are probably multistage. To comprehend mechanisms of the reactions initiated by ionizing radiation, further investigation within this area is needed.

11.5 ACKNOWLEDGMENTS

The work was performed in the frame of Young Scientists' Fund at the Faculty of Chemistry, Lodz University of Technology, Grant W-3/FMN/6G/2013.

KEYWORDS

- **Crosslink density and structure**
- **Mechanical properties**
- **Nitrile rubber**
- **Radiation crosslinking**

REFERENCES

1. Clough, R. L. (2001). High-energy radiation and polymers: A review of commercial processes and emerging applications. *Nucl. Instrum. Meth. B, 185(1–4),* 8–33.
2. Bhowmick, A. K., & Vijayabaskar, V. (2006). Electron Beam Curing of Elastomers. Rubber *Chem. Technol, 79(3),* 402–428.
3. Manaila, E., Stelescu, M. D., & Craciun, G. (2012). Advanced Elastomers – Technology, Properties and Applications. Aspects Regarding Radiation Crosslinking of Elastomers. InTech3–34.
4. Bik, J., Gluszewski, W., Rzymski, W. M., & Zagorski Z. P. (2003). EB radiation crosslinking of elastomers. *Radiat. Phys. Chem, 67(3),* 421–423.
5. Stepkowska, A., Bielinski, D. M., & Przybytniak, G. (2011). Application of Electron Beam Radiation to Modify Crosslink Structure in Rubber Vulcanizates and Its Tribological Consequences, *Acta Phys. Pol. A, 120(1),* 53–55.
6. Vijayabaskar, V., Costa, F. R., & Bhowmick, A. K. (2004). Influence of electron beam irradiation as one of the mixed crosslinking systems on the structure and properties of nitrile rubber. *Rubber Chem. Technol, 77(4),* 624–645.

7. Vijayabaskar, V., & Bhowmick, A. K. (2005). Dynamic mechanical analysis of electron beam irradiated sulfur vulcanized nitrile rubber network—some unique features. *J. Mater. Sci, 40(11)*, 2823–2831.

8. Yasin, T., Ahmed, S., Yoshii, F., & Makuuchi, K. (2002). Radiation vulcanization of acrylonitrile-butadiene rubber with polyfunctional monomers. *React. Funct. Polym, 53(2–3)*, 173–181.

9. Bik, J. M., Rzymski, W. M., Gluszewski, W., & Zagorski, Z. P. (2004). Electron Beam Crosslinking of Hydrogenated Acrylonitrile-Butadiene Rubber. *Kaut. Gummi Kunstst, 57(12)*, 651–655.

10. Stephan, M., Vijayabaskar, V., Kalaivani, S., Volke, S., Heinrich, G., Dorschner, H., Wagenknecht, U., & Bhowmick, A. K. (2007). Crosslinking of Nitrile Rubber by electron Beam Irradiation at Elevated Temperatures. *Kaut. Gummi Kunstst, 60(10)*, 542–547.

11. Vijayabaskar, V., Tikku, V. K., Bhowmick, A. K. (2006). Electron beam modification and crosslinking: Influence of nitrile and carboxyl contents and level of unsaturation on structure and properties of nitrile rubber. *Radiat. Phys. Chem., 75(7)*, 779–792.

12. Hill, D. J. T., O'Donnell, J. H., Perera, M. C. S., Pomery, P. J. (1996). An investigation of radiation-induced structural changes in nitrile rubber. *J. Polym. Sci. Pol. Chem., 34(12)*, 2439–2454.

13. Loan, L. D. (1972). Peroxide crosslinking reactions of polymers. *Pure Appl. Chem, 30(1–2)*, 173–180.

14. Zagorski, Z. P. (2002). Modification, degradation and stabilization of polymers in view of the classification of radiation spurs. *Radiat. Phys. Chem, 63(1)*, 9–19.

15. Dogadkin, B. A., Tarasova, Z. N., Golberg I. I., & Kuanyshev K. G. (1962). Effect of vulcanization structures on the strength of vulcanizates. *Kolloid. Zh, 24*, 141–151.

16. Manik, S. P., & Banerjee, S. (1969). Studies on Dicumylperoxide Vulcanization of Natural Rubber in Presence of Sulfur and Accelerators. *Rubber Chem. Technol, 42(3)*, 744–758.

17. Manik, S. P., & Banerjee S. (1970). Sulfenamide Accelerated Sulfur Vulcanization of Natural Rubber in Presence and Absence of Dicumyl Peroxide. *Rubber Chem. Technol, 43(6)*, 1311–1326.

18. Bakule, R., & Havránek, A. (1975). The dependence of dielectric properties on crosslinking density of rubbers. *J. Polym. Sci. Polym. Symp, 53(1)*, 347–356.

19. Bakule, B., Honskus, J., Nedbal, J., & Zinburg, P. (1973). Vulcanization of natural rubber by dicumyl peroxide in the presence of sulfur. *Collect. Czech. Chem. Commun, 38(2)*, 408–416.

20. Bandzierz, K., & Bielinski, D. M. (2013). Radiation methods of polymers modification: hybrid crosslinking of butadiene – acrylonitrile rubber, 244–247. Abstracts Collection on New Challenges in the European Area: Young Scientists, Baku, Azerbaijan.

21. Flory, P. J., & Rehner, J. (1943). Statistical Mechanics of Crosslinked Polymer Networks II. Swelling. *J. Chem. Phys, 11(11)*, 521–526.

22. Hwang, W.-G., Wei, K.-H., & Wu, C.-M. (2004). Mechanical, Thermal, and Barrier Properties of NBR/Organosilicate Nanocomposites. *Polym. Eng. Sci, 44(11)*, 2117–2124.

23. Saville, B., & Watson, A. A. (1967). Structural Characterization of Sulfur-Vulcanized Rubber Networks. *Rubber Chem. Technol, 40(1)*, 100–148.

24. Głuszewski, W., & Zagórski, Z. P. (2008). Radiation effects in polypropylene/polystyrene blends as the model of aromatic protection effects. *Nukleonika, 53(1)*, 21–24.
25. Seguchi, T., Tamura, K., Shimada, A., Sugimoto, M., & Kudoh, H. (2012). Mechanism of antioxidant interaction on polymer oxidation by thermal and radiation aging. *Radiat. Phys. Chem, 81(11),* 1747–1751.
26. Charlesby, A., Garratt, P. G., & Kopp, P. M. (1962). Radiation protection with sulfur and some sulfur-containing compounds. *Nature, 194*, 782.
27. Charlesby, A., Garratt, P. G., & Kopp, P. M. (1962). The use of sulfur as a protecting agent against ionizing radiations. *Int. J. Radiat. Biol, 5(5)*, 439–446.
28. Nagata, C., & Yamaguchi, T. (1978). Electronic structure of sulfur compounds and their protecting action against ionizing radiation. *Radiat. Res, 73(3)*, 430–439.

CHAPTER 12

RUBBER VULCANIZATES CONTAINING PLASMOCHEMICALLY MODIFIED FILLERS

T. GOZDEK, D. M. BIELINSKI, M. SICINSKI, H. SZYMANOWSKI, A. PIATKOWSKA, and M. STACHLEWSKA

CONTENTS

12.1　INTRODUCTION

Powders are commonly used as fillers for rubber mixes. The most popular are carbon black, silica, kaolin, or more modern like graphene, fullerenes and carbon nanotubes. The nature of their surface is the main attribute of fillers, as surface energy and specific area determine the compatibility of filler with rubber matrix and the affinity to other c ingredients. One of the major problems is the tendency of fillers to agglomeration – formation of bigger secondary structures, associated with lower level of filler dispersion, what is reflected by the decrease of mechanical properties of rubber vulcanizates [1]. Surface modification of powder can improve interaction between rubber matrix and filler. Application of low-temperature plasma treatment for this purpose has been drown increasing attention recently [2, 3].

Silica is one of the most popular mineral filler used in rubber technology. Three types of silica can be distinguished and namely: precipitated, fumed and surface-modified silica. As an amorphous material with randomly placed functional silanol groups (Fig. 12.1), it readily generates hydrogen bonds with surrounding molecules [4].

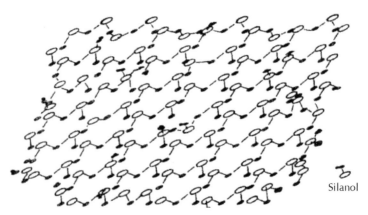

Silanol

FIGURE 12.1　Surface chemistry of silica [4].

Polar character and big specific surface area enable various modifications of silica surface. Modifying by coupling compounds is the most popular one [4]. In subject literature [5, 6] and patent declarations [7] many references on the modification processes, their kinetics, current opportunities and proposals of further development, can be found. All chemical methods have a significant disadvantage: emission of large amounts of chemical waste, usually in the form of harmful solvents.

Taking into account the necessity of their utilization, application of "clean" plasma modification has to be considered as a cost effective possibility for significant reduction of environmental hazard.

Low-temperature plasma can be generated with a discharge between electrodes in a vacuum chamber. The process used to be carried out in the presence of gas (i.e., Ar_2, O_2, N_2, methane or acetylene). Depending on the medium applied, surface of modified material can be purified, chemically activated, or grafted with various functional groups.

This chapter presents the results of low-temperature, oxygen plasma activation of silica, kaolin and wollastonite. Fillers were modified in a tumbler reactor, enabling rotation of powders in order to modify their entire volume effectively. Based on our previous work [8], the process was carried out with 100W discharge power. The time of modification varied from 8 to 64 min. Additionally, for the most favorable (in terms of changes to surface free energy) time of modification for kaolin, the process was repeated and ended with a flushing of the reactor chamber with hydrogen, in order to reduce of carboxyl groups content, generated on filler surface. Rubber mixes, filled with the modified powders, based on SBR or NBR were prepared and vulcanized. Mechanical properties of the vulcanizates were determined and explained from the point of view rubber – filler interactions and filler, estimated from micro morphology of the materials.

12.2 EXPERIMENTAL PART

12.2.1 MATERIALS

12.2.1.1 RUBBER VULCANIZATES

Three fillers were the objects of study: micro silica Arsil (Z. Ch. Rudniki S.A., Poland), kaolin KOM (Surmin-Kaolin S.A., Poland) and wollastonite Casiflux (Sibelco Specialty Minerals Europe, The Netherlands).

Rubber mixes, prepared with their application, were based on: styrene-butadiene rubber (SBR) KER 1500 (Synthos S.A., Poland) and acrylonitrile-butadiene rubber (NBR) NT 1845 (Lanxess, Germany).

Rubber mixes were prepared with a Brabender Plasticorder laboratory micro mixer (Germany), operated with 45 rpm, during 30 min. Their composition is presented in Table 12.1. The only one variable was the type of modified mineral filler (see Section 2.1).

TABLE 12.1 Composition of the Rubber Mixes Studied

Components	Content (phr)	
SBR KER1500	100	0
NBR NT1845	0	100
ZnO	5	5
Stearine	1	1
CBS	2	2
Sulfur	2	2
Arsil Silica	20	20
Modified filler	20	20

Samples were vulcanized in 160°C, time of vulcanization: 6 min (for NBR vulcanizates) and 15 min (for SBR vulcanizates).

Symbols of the prepared vulcanizates samples:
- NBR-X – composites based on NBR rubber, X – modified filler;
- SBR-X – composites based on SBR rubber, X – modified filler.

12.2.1.2 PLASMOCHEMICAL MODIFICATION OF FILLERS

Fillers studied were modified with a Diener tumbler plasma reactor (Germany). The reactor operated with the frequency of 40 kHz and the maximum discharge power of 100 W. Scheme of the reactor is shown in Fig. 12.2.

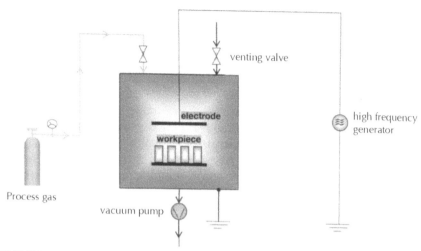

FIGURE 12.2   Scheme of the plasma reactor.

Mineral fillers were subjected to the oxygen plasma treatment during various time. Efficiency of process gas flow was 20 cm³/min, and the pressure in the reactor chamber was maintain at 30 Pa. Symbols of the modified fillers are as follows:

- A-REF – silica Arsil, virgin reference filler;
- A-XX – silica Arsil, modified during time of XX min (XX = 16; 32; 48; 64);
- K-REF – kaolin, virgin reference filler;
- K-XX – kaolin, modified during time of XX min (XX = 8; 16; 32);
- K-16H – kaolin, modified during 16 min, the process terminated with hydrogen;
- W-REF –wollastonite, virgin reference filler;
- W-XX – wollastonite, modified during time of XX min (XX = 16; 32; 48; 64).

12.2.2   TECHNIQUES

12.2.2.1   SURFACE FREE ENERGY OF FILLERS

Effectiveness of plasmochemical modification of the fillers is represented by changes to their surface free energy (SFE) and its components – polar and dispersion one. SFE was examined with a K100 MKII tensiometer (KRÜSS GmbH, Germany). Contact angle was determined using polar (water, methanol, ethanol) and non-polar (n-hexane, n-heptane) liquids. SFE and its components were calculated by the method proposed by Owens-Wendt-Rabel-Kaeble [9].

12.2.2.2   MICROMORPHOLOGY OF RUBBER

Micro morphology of rubber vulcanizates was studied with an AURIGA (Zeiss, Germany) scanning electron microscope (SEM). Secondary electron signal (SE) was used for surface imaging. Accelerating voltage of the electron beam was set to 10 keV. Samples were fractured by breaking after dipping in liquid nitrogen.

12.2.2.3   MECHANICAL PROPERTIES OF RUBBER VULCANIZATES

Mechanical properties of the vulcanizates studied were determined with a Zwick 1435 universal mechanical testing machine (Germany). Tests were carried out on "dumbbell" shape, 1.5 mm thick and 4 mm width specimens, according to PN-ISO 37:1998 standard. The following properties of the materials were determined: elongation at break (Eb), stress at elongation of 100% (SE100), 200% (SE200), 300% (SE300) and tensile strength (TS).

12.3   RESULTS AND DISCUSSION

Our previous studies revealed, that low-temperature plasma causes changes to surface free energy and its component of carbon nanotubes [10]. Plasma modification is a good method of CNT purification as an amorphous carbon is eliminated from their surface during process [11]. Purifying changes properties of CNT, and affects its dispersion in rubber matrix. It encouraged us to try plasma modification to silica, kaolin and wollastonite. The objective of the study was to characterize changes to filler surface and its susceptibility to oxygen activation, being expected to be an intermediate step in surface functionalization with various chemical groups/compounds.

12.3.1 SURFACE FREE ENERGY (SFE)

Reference silica powder represents relatively low value of surface energy and its polar component (Fig. 12.3a) – probably because of physically adsorbed water present on filler surface. After modification – regardless time of the process – SFE remains constant, however the dispersive component decreases in favor of the polar component increasing. Generally, silica remains resistant to plasma modification under the experimental conditions. However, it seems likely to change under higher discharge power.

Wollastonite behaves in a different way (Fig. 12.3b). After 48 min of plasma treatment value of its SFE reaches a maximum. Its polar component becomes almost doubled – probably because grafting of oxygen groups on filler surface. After 48 min of the treatment polar component of SFE is decreasing, probably because the process balance moves towards the surface cleaning.

Plasma treatment of kaolin (Fig. 12.3c) during 16 min results in an increase of SFE value and its polar component. After this time further changes are not observed. Hydrogen termination of the process (lasting 2 min) results in almost doubled the polar component of SFE, probably being the effect of surface present carbonyl groups reduction to the more stable carboxyl ones.

a) Surface Energy: Silica

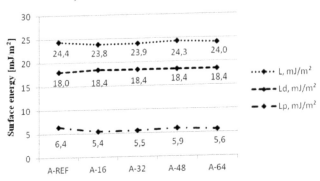

b) Surface Energy: Wollastonite

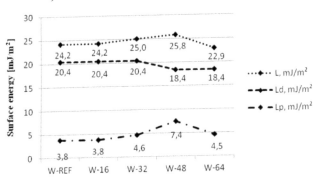

c) Surface Energy: Kaolin

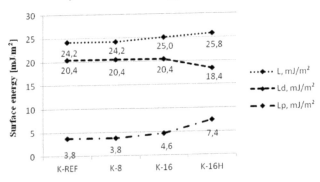

FIGURE 12.3   Results of the analysis of total surface free energy of fillers: (a) Silica, (b) Wollastonite, (c) Kaolin (L – total surface free energy, L_d – dispersive part, L_p – polar part).

12.3.2 MORPHOLOGY OF RUBBER VULCANIZATES

In order to determine the influence of rubber matrix polarity on filler dispersion and rubber-filler interaction, two kinds of rubber: NBR and SBR, were chosen. SEM pictures of the rubber vulcanizates, filled with reference and 48 min plasma treated wollastonite, are presented in Fig. 12.4. Morphology of SBR/wollastonite samples does not reveal any changes, explaining strengthening of the material (see Section 12.3.3).

Pictures of NBR-W-REF samples (Fig. 4a, b) present broken needles of wollastonite in the area of fracture, whereas in the case of NBR-W-48 sample (Fig. 4c, d) needles of wollastonite are non-broken but "pulled out" from rubber matrix. This change to morphology, reflected by lower rubber-filler interactions, responsible for worse mechanical properties of rubber vulcanizates (see Section 12.3.3), is undoubtedly the result of an increase of SFE polar component of filler after plasma treatment. The SEM pictures of the vulcanizates, no matter, containing virgin or modifies wollastonite particles, do not reveal any filler agglomeration.

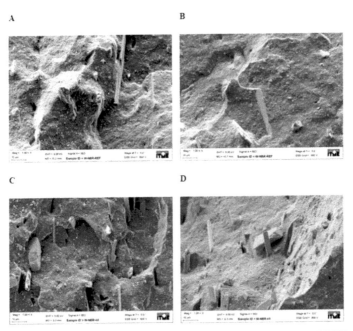

FIGURE 12.4   Morphology of rubber vulcanizates studied: A, B – NBR-W-REF; C, D – NBR-W-48; magnification 5000x.

Morphology of SBR-K-REF and SBR-K-16H samples are presented in Fig. 12.5. Agglomerates of kaolin can be seen in vulcanizate containing reference filler (Fig. 12.6a, b). Modified kaolin does not exhibit tendency to agglomeration (Fig. 12.5c, d). Better filler dispersion suggests on higher mechanical properties of the SBR vulcanizates filled with plasma treated kaolin.

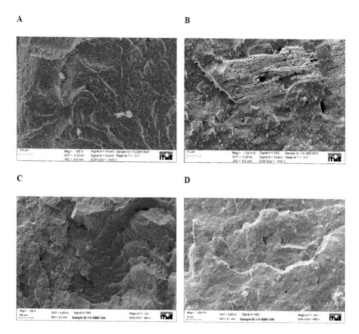

FIGURE 12.5   Morphology of rubber vulcanizates studied: A, B – SBR-K-REF; C, D – SBR-K-16H; magnification 100x (A, C) and 1000x (B, D)

12.3.3 MECHANICAL PROPERTIES OF RUBBER VULCANIZATES

Mechanical properties of the vulcanizates studied, containing virgin and plasma modified fillers are presented in Figs. 12.6a–12.6f.

Plasma modification does not cause any changes to surface free energy of silica. This is clearly reflected by the mechanical properties of silica filled rubber vulcanizates (Figs. 12.6a and b). Changes to the values of material stress at elongation 100, 200 and 300% (SE100, SE200 and SE300), its tensile strength (TS) and elongation at break (Eb), being the result of plasma treatment of the filler, are negligible.

Mechanical properties of wollastonite filled NBR vulcanizates decrease due to plasma modification of the filler (Fig. 12.6c), whereas in case of vulcanizates based on SBR an increase of TS and Eb is observed (Fig. 12.6d) – especially for the most effective 48 min treatment. SEM pictures of the vulcanizates confirm on adequate changes to their morphology.

For the rubber vulcanizates filled with kaolin (Figs. 12.6e and f), despite the biggest changes to surface free energy and its components (observed for 16 min plasma treatment followed by hydrogen termination), determined changes to mechanical properties are different in comparison to the wollastonite filled vulcanizates. The biggest increase of TS and Eb is observed for SBR/K-16H sample – about 30% as compared to the reference sample (containing virgin filler). Reinforcement of rubber seems to be dependent on overlapping effects originated from rubber-filler interactions and dispersion of filler in rubber matrix.

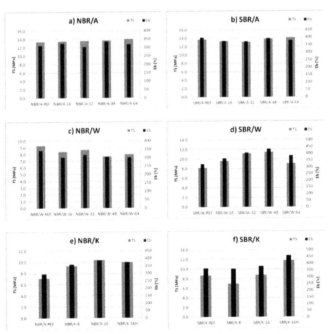

FIGURE 12.6 Mechanical properties of elastomer composites based on NBR and SBR filled with: (a–b) silica, (c–d) wollastonite, (e–f) kaolin; TS – tensile strength, Eb – elongation at break.

12.4 CONCLUSIONS

- Oxygen plasma treatment can activate surface of mineral fillers, by grafting oxygen groups on the surface.
- The efficiency of the treatment depends on the filler. Changes to surface free energy and its components are observed for kaolin and wollastonite, whereas practically no energetic effect is present in the case of silica.
- Any changes to filler particles SFE and its components effect on mechanical properties of rubber vulcanizates filled with the modi-fied filler. Improvement of mechanical properties of the materials originates increased rubber-filler interaction and better dispersion of filler particles in rubber matrix.

12.5 ACKNOWLEDGEMENT

The project was funded by the National Science Centre Poland (NCN) conferred on the basis of the decision number DEC-2012/05/B/ST8/02922.

KEYWORDS

- **Low-temperature plasma**
- **Mineral fillers**
- **Rubber vulcanizates**
- **Surface modification**

REFERENCES

1. Wolff, S. & Wang, J. (1992). Filler—Elastomer Interactions. Part IV. The Effect of the Surface Energies of Fillers on Elastomer Reinforcement. *Rubber Chemistry and Technology, 65,* 329–342.
2. Dierkes, W. K., Guo, R., Mathew, T., Tiwari, M., Datta, R. N., Talma, A. G., Noorde-meer, J. W. M., & van Ooij, W. J. (2011). A key to enhancement of compatibility and dispersion in elastomer blends. *Kautschuk Gummi Kunststoffe, 64,* 28–35.
3. Chityala, A., & van, Ooij W. J. (2000). Plasma deposition of polymer films on pmma powders using vacuum fluidization techniques. *Surface Engineering, 16,* 299–302.

4. Wang, M.-J. (1998). Effect of polymer-filler and filler-filler interactions on dynamic properties of filled vulcanizates. *Rubber Chemistry and Technology, 71,* 520–589.

5. Hair, M. L., & Hertl, W. (1971). Reaction of chlorosilanes with silica. *Journal of Physical Chemistry, 14,* 2181–2185.

6. Blume, A. (2011). Kinetics of the Silica-Silane Reaction. *Kautschuk und Gummi Kunststoffe, 4,* 38–43.

7. Revis, A. (2003). Chlorosilane blends for treating silica. *US Patent,* 6613139B1.

8. Bieliński, D., Parys, G., & Szymanowski, H. (2012). Plazmochemiczna modyfikacja powierzchni sadzy jako napełniacza mieszanek gumowych. *Przemysł Chemiczny, 91,* 1508–1512.

9. Owens, D. K., & Wendt, R. C. (1969). Estimation of the surface free energy of polymers. *Journal of Applied Polymer Science, 13,* 1741–1747.

10. Siciński, M., Bieliński, D., Gozdek, T., Piątkowska, A., Kleczewska, J., & Kwiatos, K. (2013). Kompozyty elastomerowe z dodatkiem grafenu lub MWCNT modyfikowanych plazmochemicznie. *Inżynieria Materiałowa, 6.*

11. Xu, T., Yang, J., Liu, J., & Fu, Q. (2007). Surface modification of multi-walled carbon nanotubes by O_2 plasma. *Applied Surface Science, 253,* 8945–8951.

CHAPTER 13

THE STUDY OF MODIFICATION PROCESS OF THE INDIAN RUBBER WITH FUNCTIONAL GROUPS BY OZONOLYSIS OF LATEX

L. A. VLASOVA, P. T. POLUEKTOV, S. S. NIKULIN, and V. M. MISIN

CONTENTS

ABSTRACT

This work considers the process of modification of the Indian rubber in the form of latex with ozone. Kinetics of the process has been studied. The proposed process proved to make it possible obtaining of a polymer comprised of the terminal carbonyl and carboxyl groups. It was found that an increase of ozonization degree reduced molecular mass of the Indian rubber. Main physical and mechanical properties of the rubber compounds based on the mixture of original and ozonizated rubbers were determined. Some advantages of the rubber compounds on the basis of the ozonized polymer were discussed in this chapter.

13.1   INTRODUCTION

Various functional groups involved in the macromolecules of elastomers impart all kind of rubbers a new complex of service properties as it was shown in practice. This fact considerably expands the market for rubbers as well as the areas of rubbers application.

According to a number of Russian and foreign scientists some distinctive features of the Indian rubber unlike of its nearest synthetic analog (cis-polyisoprene of different grades) are first of all connected with the absence of the functional groups in the macromolecules of synthetic isoprene rubber. The presence of such functional groups in the Indian rubber is provided by existence of non-rubber components inside, mainly of the protein type. Moreover, the difference is connected with a content and structure of gel, and besides, molecular-mass distribution of the Indian rubber macromolecules [1–4].

This viewpoint is supported by increase of the cohesion strength of the rubber compounds due to the modification of rubbers with polar compounds, for example, grafting of maleic anhydride to lithium polyisoprene [5] or by the treatment of SKI-3 rubber with n-nitrosodiphenylamine [6]. Modification of the Indian rubber is made similar to synthetic rubbers by polymer-like reactions, for example, by epoxidation [7], maleinization, hydroxylation and using some other techniques. Every of the enumerated reactions imparts a rubber some additional properties that are characteristic for the chemical properties of either functional group. The most extensive possibilities for proceeding of the chemical reactions are inherent for epoxy, as well as carbonyl groups in aldehydes and ketones because the double bond in the carbonyl groups of these compounds is strongly polarized. For example, in order to modify Indian rubber by latex epoxidation peroxyformic acid was applied (system in-situ: formic acid – hydrogen peroxide). As a result, an modification of the Indian rubber was performed in the industrial environment, that was intended for special purposes, particularly, for producing of shock-absorbing rubber soles at the railways [8].

In spite of a wide application of ozone in the laboratory routine, as well as numerous investigations of the mechanism of ozonization reaction and the study of the structure of a lot of polymer materials and chemical compounds ozone was begun to be applied in the industrial technology only recently when high-duty ozonization units have appeared in the market [9]. Now ozonolysis of unsaturated rubbers in a solution on an industrial scale is used for production of oligomers with the terminal functional groups that can be applied as the binders for a solid propellant, as the components of rubber products, tires, film-forming composites [10]. Oligoisobutylenes with ketone and carboxyl groups [11] became the first oligomers obtained in industry by ozonolysis of butyl-rubber and they were intended for the use as thickener additives to the automotive oils.

The works enumerated above are related to performing of polymer ozonolysis in the organic solvents. As a rule, the usual disadvantage of this process is a high viscosity of polymer solutions and hence, the corresponding difficulties while stirring of polymer solution with the gas mixture containing ozone. Moreover, while using ozone-air mixtures in contact with a solvent a usual release of hydrocarbon solvent from the bulk of reaction mass is observed. This can lead not only to the loss of solvent but also to the formation of flammable and explosive air-gas mixtures.

These disadvantages are completely excluded under ozonization of polymers at the stage of latex [11, 12]. The essence of elaborated technique is in the fact that latex of butadiene-styrene copolymer is treated with air-ozone mixture with a simultaneous regulating of pH value of latex within 9.5–10.5 range. In this case practically instantaneous and complete absorption of ozone by latex polymer takes place. Ozonization degree is regulated by the volume of supplied ozone-air mixture within the required mass ratio value: dry latex substance – absorbed ozone. Information on the ozonization of the Indian rubber in the form of latex is not known in scientific literature, as well as the data on the properties of the ozonizated latex and Indian rubber.

13.2   EXPERIMENTAL PART

The study of ozonization process of the Indian rubber in latex was performed with the use of the sample of non-concentrated latex of the Indian rubber received from Vietnam. Basing on the results of analyzes made by the authors latex was characterized by the following factors.

Factor	Value
Dry substance concentration, mass %	33.0
Rubber content, mass %	31.0
Content of non-rubber share, mass %.	2.9
pH value	10.6
Ammonium content calculated for NH_3, mass %	0.97
Surface tension, mN/m	46.2
Solubility of rubber in toluene, mass %	93.3
Gel content in a rubber, evolved from latex, mass %	6.1
Averaged viscous molecular mass of rubber, a.u.	1188000

Rubber for making the investigations was filled with non-staining antioxidant of phenolic type "Agidol – 2" at the rating of 0.5 mass %, after that it was released from the Indian rubber latex sample by acidifying with acetic acid with the following thorough washing of polymer with distilled water and drying it up to residual moisture content of 0.2 mass %.

The process of latex ozonization was performed with the use of pilot laboratory plant comprising of the unit applied for electrosynthesis of ozone and a reactor where interaction of the Indian rubber latex with air-ozone mixture proceeded. This pilot plant permitted to vary ozonolysis modes: to control and support pH value within the required limits, to control the process temperature, the rate of air-ozone mixture and latex supply, to control ozone concentration in the air flow in front of and after reactor, as well as the time of contact of the air-ozone mixture with latex.

The unit of ozone electrosynthesis was composed of the system for air preparation and just of ozone generator operating at 220 V 50 Hz AC and working supply voltage at the electrodes up to 6000 V. The control for ozone concentration in the air-ozone mixture was made with the use of iodometric titration. Round-bottomed flask supplied with mechanical stirrer and a bubbler for supplying of the air-ozone mixture was used as a reactor. Reactor was attached to the unit for a continuous control of pH value of the ozonizated latex and also the dosing device specifying the required value of 2% aqueous solution of potassium hydroxide for supporting of pH value in the latex within 9.5–10.5.

13.3   RESULTS AND DISCUSSION

Latex of the Indian rubber is known to have a number of specific features in its colloid-chemical properties as compared with those ones of synthetic rubber latexes. The main differences are much more sizes of the latex particles that attain 200.0–350.0 nm as well as a specific property of their protective adsorption layer consisting of a set of the natural high-molecular fatty acids, alcohols, resinous acids and protein-like compounds [13]. In this case content of the dry substances soluble in water is of about 3.0–3.5 mass % in a virgin latex, according to [13]. This is approximately in accordance with the result of analysis of the investigated latex (2.9 mass %). Fatty acids in latex are presented by oleic, linoleic ones as well as by other carboxyl-containing compounds that have concentration of 1.05–2.05 mass % in the acetone extract for Indian rubber latex. Content of the protein substances in the freshly gathered latex was up to 4.0 mass %, while free amino acids were found in the concentration of 0.2 mass %.

This rather complicated composition of compounds participating in the formation of adsorption layer on the globules of Indian rubber latex provides an ability of its existence in a dependence of pH value in anion (pH>8.0) or in cation (pH<5.0) form keeping to some extent its aggregative stability. As a rule, Indian rubber latex is filled with ammonia in order to stabilize it and is characterized by pH = 10.0–10.5. Acidifying of latex during its storage or under its targeted filling with acids, for example, acetic acid, results in latex coagulation. This specific feature was taken into account while refining ozonization conditions since under latex ozonization they are somewhat acidified due to the formation of carboxylic groups in polymer. In this case pH is reduced up to the neutral or even low-acid value that is capable to cause an untimely latex coagulation.

Alkaline reaction leads to the formation of the corresponding fatty-acid salts (anion-active surface-active substances), adsorbed on the surface of polymer particles. Surface tension of the investigated latex determined by tensometric method was of 46.2 mN/m thus defining an extra aggregative stability of Indian rubber latex during air-ozone bubbling.

Experimentally determined reaction rate constant of ozone interaction with -C=C- bonds in the polymer of Indian rubber latex are of the order of 2–6×10⁴ L/mole·s. Therefore, reaction rate of the unsaturated polymer of isoprene is mainly limited by supply rate of ozone into reactor.

Process of Indian rubber latex ozonization was made in the following way. Latex in the amount necessary for making the experiment was supplied to the round-bottom flask with a mechanical stirrer and after its activation aqueous emulsion of antifoam agent was added. Next, air-ozone mixture with the concentration of 16–18 mg/dm³ was also supplied through the bubbler. The temperature of the process was supported within 18–20 °C, while pH value was within 9.5–10.5. Quantitative absorption of ozone with latex was controlled by iodometric analysis of the outgoing air evolving from the reactor flask through absorbing 2% solution of potassium iodide.

In order to reveal the changes taking place with Indian rubber latex and the rubber in the process of ozonization we made the investigations of ozone absorption with latex basing on the estimation of 0.3; 0.7; 0.8; 1.0; 1.5; 4.0; 8.0 mass % of ozone absorption relative to the rubber. An increase of ozonization degree did not result in any complications in the process connected with latex aggregative stability.

Results of analysis of the main colloid-chemical properties for the modified and original latexes and Indian rubber are presented in Table 13.1.

TABLE 13.1 The Change of Colloid-Chemical Properties of Latex and Indian Rubber in the Process of Ozonization

No	Amount of the bound ozone relative to the rubber, mass %	Solid residue, mass %	pH value after ozonization	Surface tension, mN/m	Content of carbonyl groups in rubber, mass %	Content of carboxyl groups in rubber, mass %	Solubility of rubber in toluene, mass %	Gel content, mass %	Averaged-viscous molecular mass (MM) of rubbers released from rubber samples
1	2	3	4	5	6	7	8	9	10
1	0	33.00	10.70	46.20	not determined	not determined	93.90	6.10	1188000
2	0.30	32.80	10.50	42.50	0.11	0.10	93.00	6.90	1082100
3	0.70	32.90	10.40	41.40	0.30	0.32	93.60	6.40	997274
4	0.80	32.80	10.40	40.50	0.40	0.40	92.30	7.70	990120
5	1.00	33.20	10.30	38.90	0.50	0.41	91.70	8.30	993000
6	1.50	31.90	10.10	36.00	0.71	0.70	89.50	10.50	825037
7	4.00	33.00	9.90	35.00	2.00	1.80	85.00	15.00	394094
8	9.00	32.50	8.70	35.00	4.50	4.10	83.70	16.30	138943

Note: for the Indian rubber, separated from the given latex sample, stabilized with anti-oxidant Agidol-2 (0.5 mass %) after plasticization with rolling mills at the temperature of 25–30 °C and the gap of 2 mm for 10 min Mooney viscosity was reduced up to 50.0–55.0 units.

As it is seen from the obtained data, the ozonization process of the Indian rubber latex with an increase of the amount of the bound ozone is entailed by the regular decrease of the surface tension value from 46.0 mN/m for the original latex to 35.0 mN/m for the ozonizated latexes. Consequently, the degree of adsorption saturation for the surface of latex globules increased from 55–60 % to 100 %.

This phenomenon is connected with the formation of intermediate ozonide cycles under ozonolysis of the binary bonds and dissociation of these cycles up to the fragments of macromolecules with carboxyl and carbonyl terminal groups. Then carboxyl groups in the ionized form (pH – 9.5–10.5) can additionally participate in the formation of protective adsorption layer on the surface of latex particles. Due to increase of the adsorption saturation Indian rubber latex gained the enhanced aggregative stability that was observed after extraction of rubber samples from ozonizated latexes by coagulation method. It should be noted that latex concentration in the process of its treatment with ozone was not actually changed and it was not accompanied by formation of coagulum; this fact confirmed technological ability of the considered way to modify Indian rubber.

While investigating of the properties of ozonizated rubbers a continuous decrease of the averaged-viscosity molecular mass was observed with an increase of the ozone amount attached to the rubber. At the same time the content of gel fraction remained almost the same and slightly increased with the rise of ozonization up to 4–8 mass % and, consequently, with the increase of the number of functional groups in the polymer chain.

It was found that the value of mean molecular weight of original Indian rubber present in the fraction soluble in toluene was equal approximately to 1,188,000 units. Ozonization of the Indian rubber in latex with the attachment of 0.8–1.0 mass % of ozone reduced mean molecular weight up to 990,000–993,000 units, that is, to the value that approximately corresponded to the working viscosity of mechanically plasticized rubber with Mooney viscosity equal to 50–55 units. Dependence of the change of molecular mass for the ozonizated rubber on the amount of the bound ozone is presented in Table 13.1. In all of the samples of modified polymers extracted from the ozonizated Indian rubber latex content of the carbonyl and carboxyl functional groups regularly increased with an increase of amount of the attached ozone. The presence of carbonyl groups related to aldehyde or ketone groups in the ozonizated rubbers was confirmed by the results of IR spectroscopy in accordance with absorption band in the range of 1710–1740 cm^{-1} and by UV spectroscopy by the presence of absorption at 270–280 nm. The amount of carboxyl groups was determined by acid-base titration of rubber solution in toluene with 0.1 N alcohol solution of potassium hydroxide. It should be noted that according to the known scheme of ozonization reaction functional groups incorporated into the Indian rubber can be arranged at the ends of molecular chain. It is connected with the fact that Indian rubber has cis-1,4-structure and actually does not involve the links of 1,2- or 3,4-attachments. Therefore with a release of internal binary bonds under the effect of ozone Indian rubber can form only terminal functional groups. This feature in the arrangement of functional groups in the ozonizated polymers can play an essential positive role since their participation in the chemical reactions with the components of rubber compounds facilitates elongation of molecular chains and it will not lead to scorching. Similar disadvantage is characteristic of synthetic carboxylate rubbers, which have statistical distribution of the carboxylic groups along the copolymer chain.

To estimate the properties of ozonizated Indian rubber in the rubber compound a sample of 0.4 kg mass was obtained with ozonization degree of 0.8 mass % (sample № 6 in Table 13.1). This sample was used for the preparation of the rubber compound according to the standard formulation approved in tire production. Rubber compound prepared on the basis of the rubber extracted from the sample of original latex that was preliminarily plasticized with the use of cold rolling mills with the gap of 2 mm for 3 min was applied as the reference sample. The choice of plasticization time was connected with the necessity of decrease of the rubber molecular mass up to the value of that one obtained from the ozonizated latex.

Rubber compounds were vulcanized at 133 °C for 30 min. Results of the tests are presented in Table 13.2.

TABLE 13.2 Properties of the Rubber Compounds on the Basis of Initial and Ozonizated Indian Rubbers

No	Name of the factors	Initial rubber	Ozonizated rubber
1	2	3	4
1	Mooney viscosity, MB 1+4 (100°C) rubber compound	18.00	22.00
2	Cohesion strength of rubber compound, kg-force/cm^2	3.90	7.21
3	Strength of bond with metal according to GOST 209–62, kg-force/cm^2	15.50	15.20
4	Tension under 300% elongation, MPa	7.00	8.40
5	Conditional strength under rupture, MPa	31.50	33.40
6	Relative elongation under rupture, %	670.00	625.00
7	Relative residual deformation after rupture, %	28.00	26.00

Note: factors of NN 3–7 are presented for vulcanizates.

13.4 CONCLUSIONS

Analyzing the obtained experimental data one can conclude that rubber ozonization did not make worse the basic physicomechanical quality factors of the rubbers made on its basis. At the same time cohesion strength of the raw rubber considerably increased and the strength of the vulcanized rubber somewhat increased as well. It should be especially noted that the tested sample of ozonizated Indian rubber does not require plasticization with the use of roll mills, that is rather power- and labor-consuming operation accepted in the technology of tire production.

KEYWORDS

- **Indian rubber**
- **Latex**
- **Ozonization**
- **Physicomechanical properties**

REFERENCES

1. Garmonov, I. V. (1973). *Caoutchouc and Rubber*, 5, 6–15.
2. Poddubnyj, I. Ya., Grechanovskii, V. A., & Ivanova, L. S. (1972). Molecular structure and microscopic properties of synthetic cis-polyisoprene. Report at the international symposium on isoprene rubber. Moscow, 20–24/IX, 1972. M.: TsNIITEneftekhim, 19.
3. Briston, J. M., Canin, J. I., & Mullins, L. (1972). Comparison of the properties and performance specifications of Indian rubber and synthetic cis-polyisoprene. Report at the international symposium on isoprene rubber. Moscow, 20–24/IX, 1972. M.: TsNIITEneftekhim, 40.
4. Greg, E. S., & McKey, J. H. (1972). Differences of technological properties of synthetic polyisoprene and Indian rubber. Influence of non-rubber components. Report at the international symposium on isoprene rubber. Moscow, 20–24/IX, 1972. M.: TsNIITEneftekhim, 35.
5. Sobolev, V. M., & Borodina, I. V. (1977). Industrial synthetic rubbers. M.: *Khimia*, 256.
6. Lykin, A. S., et al. (1972). Investigations of n-nitrosodiphenylamine effect on cohesion strength and stability of the coating rubbers made from SKI-3. Report at the international symposium on isoprene rubber. Moscow, 20–24/IX, 1972. M.: TsNIITEneftekhim, 18.

CHAPTER 14

INFLUENCE OF THE STRUCTURE OF POLYMER MATERIAL ON MODIFICATION OF THE SURFACE LAYER OF IRON COUNTERFACE IN TRIBOLOGICAL CONTACT

DARIUSZ M. BIELIŃSKI, MARIUSZ SICIŃSKI, JACEK GRAMS, and MICHAŁ WIATROWSKI

CONTENTS

ABSTRACT

The degree of modification of the surface layer of Armco iron by sulfur, produced by sliding friction of the metal sample against: ebonite, sulfur vulcanizate of styrene–butadiene rubber (SBR), polysulphone or polysulfide rubber, was studied. Time of Flight Secondary Ion Mass Spectroscopy (TOF–SIMS) and confocal Raman microscopy techniques, both confirmed on the presence of iron sulfide (FeS) in the surface layer of metal counterface after tribological contact with SBR or polysulfide rubber. For the friction couple iron–ebonite, the presence of FeS was confirmed only by TOF–SIMS spectra. FT–Raman analysis indicated only on some oxides and unidentified hydrocarbon fragments being present. Any sulfur containing species were not found in the surface layer of iron counterface due to friction of the metal against polysulphone. The degree of iron modification is determined by the loading of friction couple, but also depends on the way sulfur is bonded in polymer material. Possibility for modification is limited only to materials, which contain sulfur either in a form of ionic sulfide crosslinks (SBR and ebonite) or side chains (polysulfide rubber). Degradation of polymer macromolecules during friction (polysulphone and ebonite – in this case under high loading) does not lead to the formation of FeS. Chemical reaction between sulfur and iron takes place only in the case of ionic products of polymer destruction containing sulfur.

14.1 INTRODUCTION

Interest towards chemical reactions accompanying friction has been growing since last years. This is reflected by significant progress in very important area of tribology, called tribochemistry [1]. One of its priorities are studies on chemical reactions taking place in the surface layer of materials constituting the friction couple and their exploitation consequences, e.g. concerning creation of protective layers, lowering wear, etc.

An increase of temperature in tribological contact during friction is well known. It facilitates the phenomenon of selective transfer of polymer components, followed by their chemical reaction with the surface layer of metal counterface, in the case of rubber–metal friction couple. The modification can not only effect composition and structure of the surface layer of polymer but metal as well [2]. Our previous studies confirmed on the possibility to modify the surface layer of iron counterface by sliding friction against sulfur vulcanizates of styrene–butadiene rubber [3, 4]. Extend of modification is related to the kind of dominated sulfur crosslinks. An increase of temperature accompanying friction facilities breaking of crosslinks present in vulcanizate, especially polysulfide ones [5]. The highest degree of modification was detected for Armco iron specimen working in tribological contact with rubber crosslinked by an effective sulfur system of short: mono- and di- to long polysulfide crosslinks ratio equal to 0.55. Polysulfide crosslinks characterize themselves by the lowest energy from the range of crosslinks created during conventional sulfur vulcanization (-C-C-, -C-S-C-, -C-S$_2$-C-, -C-S$_n$-C-; where $n \geq 3$) [6]. So, their breaking as first is the most probable. As a result, the release of sulfur ions, representing high chemical reactivity to iron, takes place. FeS layer of 100–150 nm thickness was detected on iron specimen subjected to friction against sulfur vulcanizates of SBR [3]. It lubricates efficiently the surface of metal, reducing the coefficient of friction [7]. As a compound of low shearing resistance, FeS is easily spreaded in the friction zone, adheres to metal counterface, penetrating its microroughness. Even very thin layer of FeS showed to be effective due to high adhesion to iron. Metal oxides (mainly Fe$_3$O$_4$) being created simultaneously on the metal surface, act synergistically with FeS, making it wear resistance significantly increased [8]. This paper is to compare other polymer materials containing sulfur to conventional rubber vulcanizates in terms of their ability to the surface modification of iron. The polymers studied vary from sulfur vulcanizates either according to crosslink density (ebonite) or the kind of sulfur incorporation in macromolecules (polysulphone – rigid material and polysulfide rubber – elastomer).

14.2 EXPERIMENTAL PART

14.2.1 MATERIALS

Surface polished specimen made of Armco iron were subjected to extensive friction against:
- polysulphone PSU 1000 (Quadrant PP, Belgium),
- crosslinked polysulfide rubber LP-23 (Toray, Japan),
- ebonite based on natural rubber [9], or

- carbon black filled sulfur vulcanizate of styrene-butadiene rubber Ker 1500 (Z. Chem. Dwory, Poland).

Composition of the materials studied is given in Table 14.1.

TABLE 14.1　Composition of the Polymer Materials Studied

Material Components	Polysulfide rubber	Ebonite	Conventional rubber (SBR)	Polysulphone
Styrene-butadiene rubber, Ker 1500			100	
Natural rubber, RSS II		100		
Polysulfide rubber, LP-23	100			
Poly(sulphone), PSU 1000				100
Stearic acid			1	
Zinc oxide, ZnO			3	
HAF carbon black, Corax N 326			50	
Ebonite powder		50		
Linseed oil		2		
Isostearic acid	0.20			
Manganese dioxide, MnO_2	10			
Tetramethylthiuram disulfide, TMTD	0.50			
N-third buthyl-di-benzothiazolilosulphenamide, TBBS			2.20	
Zinc dithiocarbamate, Vulkacit		1		
Sulfur, S_8		42	0.80	

Rubber mixes were prepared with a David Bridge (UK) roller mixer. Specimen for further examinations were vulcanized in a steel mold at 160 °C, during time $\tau_{0.9}$, determined rheometrically with a WG 05 instrument (Metalchem, Poland), according to ISO 3417. Liquid polysulfide rubber was cured at room temperature by means of chemical initiator, activated by MnO_2. Polysulphone specimen were prepared by cutting off from a rod.

Modification of Armco iron counterface was performed by rubbing of polymer materials studied against metal specimen. The process was realized with a T-05 tribometer (IteE-PIB, Poland).

14.2.2 TECHNIQUES

14.2.2.1 TOF-SIMS

Studies were carried out by means of an ION-TOF SIMS IV instrument (Germany), operating with a pulse ^{69}Ga ion gun of beam energy 25 kV. Primary ion dose was any time kept below 3×10^{11} cm^{-2} (static mode). Negative and positive ion spectra of iron specimen were collected in the range of m/z 1–800, before and after friction against polymer materials. Analysis was narrowed to the range of m/z<35, which showed to be the most relevant in terms of sulfur modification. The most informative signals can be subscribed to: H- (m/z = 1), C- (m/z = 12), CH- (m/z = 13), O- (m/z = 16), OH- (m/z = 17), C_2- (m/z = 24), S- (m/z = 32) and SH- (m/z = 33). Any time counts of S- and SH- were normalized to the total number of counts present in the spectrum.

14.2.2.2 RAMAN SPECTROSCOPY

Studies were carried out by means of a Jobin – Yvon T64000 (France) instrument, operating with a laser of 514.5 nm line and power of 50–100 W. The spectrometer was coupled with a BX40 Olympus confocal microscope, operating with Olympus LMPlanFI 50× (NA=0.50) or LMPlan 50× (NA=0.75) objectives. The surface of iron specimen was examined with acquisition time of 240–540 s, at least in two distant places, before and after friction. The Internet RASMIN database [10] was used for material identification. Characteristic absorption bands for FeS are present at wavelengths of 270 and 520 cm^{-1} (Fig. 14.1).

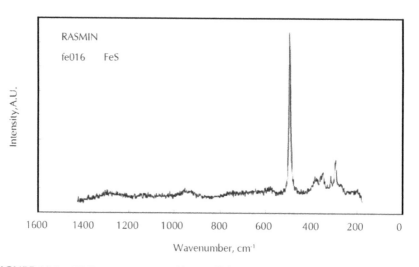

FIGURE 14.1   FT-Raman spectrum of iron sulfide [10].

14.2.2.3   TRIBOLOGICAL CHARACTERISTICS

Tribological characteristics of the materials studied were determined with a T-05 (ITeE – PIB, Poland) tribometer, operating with a block-on-ring friction couple. Ring made of polymer material was rotating over a still block made of Armco iron. The instrument worked together with a SPI-DER 8 Hottinger Messtechnik (Germany) electronic system for data acquisition. The way for data analysis has been described in our previous work [11]. Polymer rings of diameter 35 mm, rotating with a speed of 60 rpm were loaded within the range of 5–100 N, during 60–120 min.

14.3   RESULTS AND DISCUSSION

In order to explain the influence of the way sulfur is bonding in polymer materials on modification of the surface layer of iron, the metal counterface was subjected to sliding friction against:
- polysulphone (load of 20 N/ time 2 h),
- ebonite (load either 20 or 100 N/ time 2 h),
- SBR vulcanizate (load 20 N/ time 2 h), and
- polysulfide rubber (load 5 N/ time 1 h).

The load and time of friction in the last case have to be decreased due to low mechanical strength of polysulfide rubber.

From the specific spectra of secondary ions (Fig. 14.2) and comparison between normalized counts for particular cases (Fig. 14.3) it follows, that the highest amount of sulfur, in a form of SH- ions, was transferred to the surface layer of iron counterface by ebonite. In the case of polysulphone, due to strong sulfur bonding to macromolecular backbone (Fig. 14.4) and different from other polymers studied mechanisms of mechano-degradation, the expected effect of sulfur transfer is practically absent.

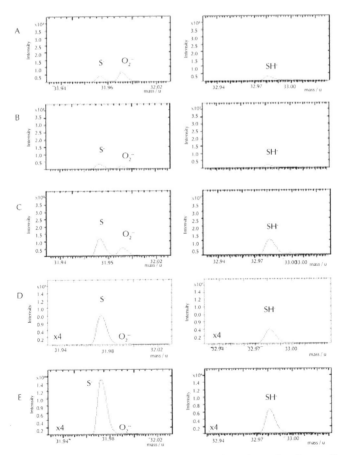

FIGURE 14.2 Specific TOF-SIMS spectra collected from the surface layer of iron Armco specimen, subjected to friction against various polymer materials studied: A – virgin, B – polysulphone, C – SBR vulcanizate, D – ebonite/100 N, E – ebonite/20 N.

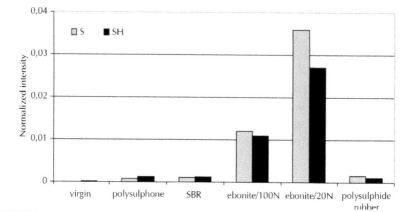

FIGURE 14.3 Normalized TOF-SIMS spectra counts of S- and SH- ions for the surface layer of Armco iron specimen subjected to friction against various polymer materials studied.

FIGURE 14.4 Chemical structure of polysulphone.

The amount of iron sulfide, created in the surface layer of iron counterface depends on reactivity of sulfur containing polymer fragment being released during friction and their concentration in the friction zone. From possible substrates, involved in the creation of FeS, the highest affinity to iron exhibit polysulfide crosslinks and ionic products of their destruction, released from some polymer materials subjected to intensive friction against the metal counterface. They can be produced only in the case of SBR vulcanizate and ebonite, what can be explained by their chemical structure. One should pay attention to different load being applied for the polymer materials studied. In the case of unfilled polysulfide rubber, the time of friction has additionally to be limited due to low mechanical strength of the material. However, an example of ebonite, demonstrates that an increase of loading not necessarily has to lead to higher extent of modification of iron counterface during friction (Figs. 14.2 and 14.3).

It can be the result of prevailing, under extreme friction conditions, radical degradation of macromolecules, accompanied by intensive oxidation of polymer. Distribution of all ions in the surface layer of metal is uniform.

Apart sulfur ones, released from polymer materials, also oxides are formed, what facilitates antifriction properties of metal [12]. Iron sulfide present can further be oxidized during friction to sulphones, which are even better lubricating agents. However, under too high loading conditions ionic mechanism of crosslink breaking is not able to show up, loosing to macromolecular degradation, what explains lower efficiency of the modification of iron with sulfur. In order to confirm TOF–SIMS data on FeS presence in the surface layer of Armco iron subjected to friction against various polymer materials, complementary studies with Raman spectroscopy were carried out. Comparing collected spectra (Fig. 14.5) to the standard FeS spectrum from the database (Fig. 14.1), only the spectra of iron surface after friction against SBR vulcanizate and polysulfide rubber indicate on possible sulfur modification. Analysis of the surface of Armco iron specimen subjected to friction against polysulphone or ebonite did not bring unique results. FT–Raman spectra contain signals the most likely coming from degraded fragments of macromolecules or unidentified compounds containing carbon, oxygen and hydrogen. For these two cases any absorption peak in the region characteristic for the FeS could not be assigned.

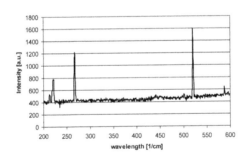

A – polysulphide rubber

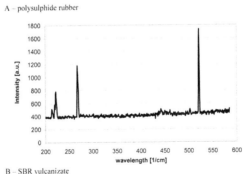

B – SBR vulcanizate

FIGURE 14.5   *(Continued)*

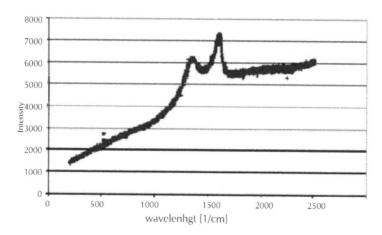

C – ebonite/100 N

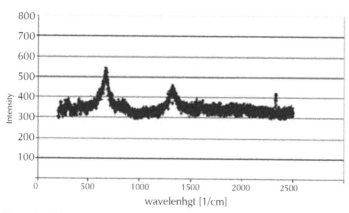

D – polysulphone

FIGURE 14.5   FT–Raman spectra of the surface layer of Armco iron specimen subjected to friction against various polymer materials studied. (A) polysulfide rubber; (B) SBR vulcanizate; (C) ebonite/100 N; (D) polysulphone.

Tribological characteristics of polymer materials sliding against Armco iron are demonstrated in Fig. 14.6. Friction force and energy curves vary from material to material due to their different chemical structure and related mechanical properties. From tribological point of view, the most efficient modification has to be subscribed to ebonite – iron friction couple. In this case median of the friction force and discrete levels of energy exhibit the most stable courses among all the friction couples studied. The second run, repeated for new ebonite roll after 2 h of previous frictional modification of Armco iron, resulted in >10% reduction of the coefficient of friction (Fig. 14.6B). The first run for SBR vulcanizate (Fig. 14.6C) exhibits "classical" course with characteristic maximum of the coefficient of friction, which appears already after some minutes from the start of experiment. For the first 100 min value of the friction force gradually decreased from 38 down to 26 N, eventually stabilizing at this level. In the second run, the friction force come back to the initial value, but right after beginning immediately goes down to the final value after the first run. It means, that in the case of SBR vulcanizates, the modification of metal counterface is the most important for the beginning of friction. Tribological characteristics determined for polysulfide rubber (Figs. 14.6 E and 14.6F) are not so stable as for ebonite or SBR vulcanizates, probably because of poor mechanical properties of polysulfide rubber. Nevertheless, the modification of the surface layer of Armco iron, confirmed by TOF – SIMS and Raman spectroscopy, is also reflected by tribological data. In the first cycle, the friction force is maintained at the level of 17–18 N during the first 25 min, and suddenly goes down to 10 N, which level is kept constant till the end of experiment. The drop is reflected by significant increase of the energy component responsible for high energy vibrations (200–600 Hz). The vibrations of such energy are not present in tribological characteristics of elastomers [11]. Similarly to SBR vulcanizate, the second run for polysulfide rubber starts from higher value of the friction force, which quickly goes down and stabilizes itself at the final level of the first run. In the case of polysulphone (Fig. 14.6G and 14.6H) any tribological effects, able to be subscribed to the surface modification of iron counterface, have not been observed. During the first run, value of the friction force is increasing, eventually reaching stabilization at the level of 9 N, shortly before the end of experiment. The second run starts from the friction force value of 5 N, which gradually increases up to the final level of the first run. At this moment, which requires about 40 min from start, its course become very unstable, probably because of intensive wear of iron counterface influencing experimental data.

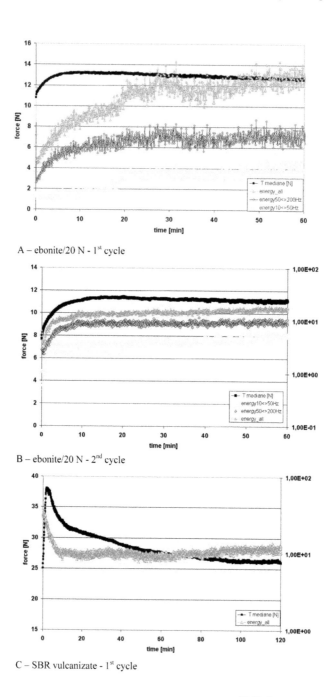

A – ebonite/20 N - 1st cycle

B – ebonite/20 N - 2nd cycle

C – SBR vulcanizate - 1st cycle

FIGURE 14.6   *(Continued)*

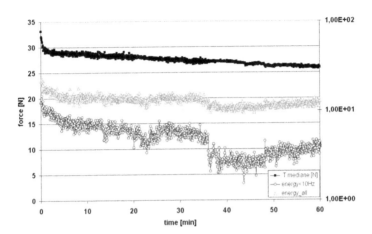

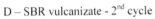

D – SBR vulcanizate - 2nd cycle

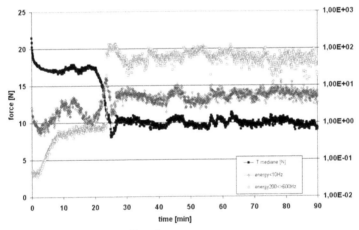

E – polysulphide rubber - 1st cycle

FIGURE 14.6 *(Continued)*

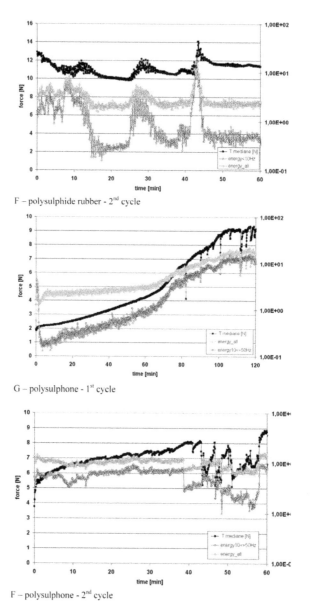

F – polysulphide rubber - 2nd cycle

G – polysulphone - 1st cycle

F – polysulphone - 2nd cycle

FIGURE 14.6   Tribological characteristics of Armco iron – polymer material friction couple. A – ebonite/20 N – 1st cycle; B – ebonite/20 N – 2nd cycle; C – SBR vulcanizate – 1st cycle; D – SBR vulcanizate – 2nd cycle; E – polysulfide rubber – 1st cycle; F – polysulfide rubber – 2nd cycle; G – polysulphone – 1st cycle; F – polysulphone – 2nd cycle.

14.4 CONCLUSIONS

- Extensive friction against polymer materials containing sulfur can result in surface modification of Armco iron counterface with FeS. Extent of the modification depends on the way of sulfur bonding to macromolecules.
- The results obtained point of higher efficiency of modification, when sulfur is present in ionic crosslinks of polymer material – SBR vulcanizate and ebonite, contrary to constituting its macromolecules – polysulphone. Degradation of the latter is of radical character and its products immediately react with atmospherics oxygen. Only ionic species, produced by breaking of sulfide crosslinks, are able to react with iron.
- The extent of modification is straightly related to the amount of sulfur being present in crosslinks. The highest amount of sulfur in the case of ebonite results in the highest degree of modification of the surface layer of Armco iron after tribological contact. Application of cured polysulfide rubber is less effective due to the lack of sulfide crosslinks in structure of the material. Additionally, its poor mechanical properties are responsible for a transfer of low molecular weight products of wear onto the metal counterface, whereas "strong" polysulphone makes Armco iron sample worn off.
- TOF-SIMS data for Armco iron point on the highest degree of sulfur modification being the result of friction against ebonite, SBR vulcanizate and cured polysulfide rubber. The spectra represent the highest amount of species containing sulfur.
- Tribological characteristics confirm the influence of sulfur modification of metal counterface on lowering friction for the metal–polymer couples studied. In the case of ebonite, the coefficient of friction reduced significantly for the whole experimental run, whereas application of SBR vulcanizate or polysulfide rubber was effective only during the first period of experimental cycles. Any improvement of tribological characteristics was not assigned for polysulphone. The polymer was observed to worn the surface of iron counterface, what resulted in increase of the coefficient of friction in this case.

KEYWORDS

- Iron
- Polymers
- Sliding friction
- Surface modification

REFERENCES

1. Płaza, S. (1997). *Physics and Chemistry of Tribological Processes*, University of Łódź Press, Łódź.
2. Rymuza, Z. (1986). *Tribology of Sliding Polymers*, WNT, Warsaw.
3. Bieliński, D. M., Grams, J., Paryjczak, T., & Wiatrowski, M. (2006). Tribological modification of metal counterface by rubber, *Tribological Letters, 24*, 115–118.
4. Bieliński, D. M., Siciński, M., Grams, J., & Wiatrowski, M. (2007). Influence of the crosslink structure in rubber on the degree of modification of the surface layer of iron in elastomer-metal friction pair, *Tribologia, 212*, 55–64.
5. Boochathum, P., & Prajudtake, W. (2001). Vulcanization of cis- and trans-polyisoprene and their blends: cure characteristics and crosslink distribution, *Eur. Polym, J. 37*, 417–427.
6. Morrison, N. J., & Porter, M. (1983). Temperature effects on structure and properties during vulcanization and service of sulfur-crosslinked rubbers, *Plast. Rubber Proc. Appl, 3*, 295–304.
7. Grossiord, C., Martin, J. M., Th Mogne, Le., & Palermo, Th. (1998). In situ MoS formation and selective transfer from MoDPT films, *Surf. Coat. Technol, 108–109*, 352–359.
8. Wang, H., Xu, B., Liu, J., & Zhuang, D. (2005). Investigation on friction and wear behaviors of FeS films on L6 steel surface, *Appl. Surf. Sci, 252*, 1084–1091.
9. Gaczyński, R. ed. (1981). *Rubber. Handbook for Engineers and Technicians*, WNT, Warsaw, Tab. IV-*35*, 323.
10. http: www.aist.go.jp/RIOBD/rasmin/E-index.htm
11. Głąb, P., & Bieliński, D. M. (2004). Maciejewska, K. An attempt to analysis of stick-slip phenomenon for elastomers, *Tribologia, 197*, 43–50.
12. Wang, H., Xu, B., Liu, J., & Zhuang, D. (2005). Characterization and tribological properties of plasma sprayed FeS solid lubrication coatings, *Mater. Characterization, 55*, 43–49.

PART 4
STABILITY OF ELASTOMERS

CHAPTER 15

A RESEARCH NOTE ON MORPHOLOGY AND STABILITY OF POLYHYDROXYBUTYRATE ELECTROSPUN NANOFIBERS

A. A. OLKHOV, O. V. STAROVEROVA, A. L. IORDANSKII and G. E. ZAIKOV

CONTENTS

ABSTRACT

This research note focuses on process characteristics of polymer solutions, such as viscosity and electrical conductivity, as well as the parameters of electrospinning using poly-3-hydroxybutyrate modified by titanium dioxide nanoparticles, which have been optimized. The structure of materials has been examined by means of X-ray diffraction, differential scanning calorimetry, IR-spectroscopy, and physical-mechanical testing.

15.1   INTRODUCTION

This research work focuses on process characteristics of polymer solutions, such as viscosity and electrical conductivity, as well as the parameters of electrospinning using poly-3-hydroxybutyrate modified by titanium dioxide nanoparticles, which have been optimized. Both physical-mechanical characteristics and photooxidation stability of materials have been improved. The structure of materials has been examined by means of X-ray diffraction, differential scanning calorimetry (DSC), IR-spectroscopy, and physical-mechanical testing. The fibrous materials obtained can find a wide application in medicine and filtration techniques as scaffolds for cell growth, filters for body fluids and gas-air media, and sorbents.

Titanium dioxide nanoparticles are the most attractive because of the developed surface of titanium dioxide, the formation of surface hydroxyl groups with high reactivity resulted from reacting with electrolytes as crystallite sizes decrease down to 100Å and lower, and a high efficiency of oxidation of virtually any organic substance or many biological objects.

Many modern applications of TiO_2 are based on using its anatase modification, which shows the minimum surface energy and greater concentration of OH-groups on sample surface compared with other modifications. According to Dadachov's patents [1, 2], the nanosized η-modification of TiO_2 is considerably superior to anatase in the above mentioned properties.

Polyhydroxybutyrate (PHB) is the most common type of a new class of biodegradable termoplasts, namely polyoxyalkanoates. It demonstrates a high strength and the ability to biodegrade under natural environmental conditions, as well as a moderate hydrophilicity and nontoxicity (biodegrades to CO_2 and water) [3]. PHB shows a wide range of useful performance characteristics; it is superior to polyesters which are the standard materials for implants, can find application in different branches of medicine, and is of great importance for cell engineering due to its biocompatibility [4]. However, its strength and other characteristics, such as thermal stability, gas permeability, and both reduced solubility and fire resistance, are insufficient for its large-scale application.

The objective of the research was to prepare ultra-fine polymer composition fibers based on polyhydroxybutyrate and titanium dioxide and to determine the role played by nanosized titanium dioxide modifications in achieving special properties of the compositions.

15.2   EXPERIMENTAL PART

The nanosized η-TiO_2 and anatase (S12 and S30) were prepared by sulfate process from the two starting reagents, $(TiO)SO_4 \cdot xH_2SO_4 \cdot yH_2O$ (I) and $(TiO)SO_4 \cdot 2H_2O$ (II) correspondingly [5].

Samples of titanium dioxide and its compositions with polymers were analyzed by X-ray diffraction technique, using HZG-4 (Ni filter) and (plane graphite monochromator) diffractometers, CuK_α radiation, diffracted beam, in the range of 2θ 2–80°, rotating sample, stepwise mode (the impulse accumulation time is 10 s, by step of 0.02°). Experimental data array was processed with PROFILE FITTING V 4.0 software. Qualitative phase analysis of samples was carried out by using JCPDS PDF-2 database, ICSD structure data bank, and original papers.

Particle sizes (coherent scattering region) of TiO_2 samples were calculated by the Selyakov-Scherrer equation $L = \dfrac{k\lambda}{\beta \cdot \cos\theta_{hkl}}$, where $\beta = \sqrt{B^2 - b^2}$ is physical peak width for the phase under study (diffraction reflections were approximated by Gaussian function), B is integral peak width, b is instrument error correction ($b \sim 0.14°$ for α-Al_2O_3 as a reference), $k \sim 0.9$ is empirical coefficient, λ is wavelength. Calculations were based on a strongest reflection at $2\theta \sim 25°$. Standard deviation was ±5%.

Starting PHB with molecular weight of 450 kDa was prepared through microbiological synthesis by BIOMER (Germany). Chloroform (CFM) was used as solvent for preparing polymer solution. Both HCOOH (FA) and $[CH_3(CH_2)_3]_4N$ (TBAI) were used as special additives.

Electrostatic spinning of fibers based on PHB and titanium dioxide was carried out with original laboratory installation [6].

The dynamic viscosity of polymer solutions of various compositions was measured as a function of PHB concentration with Heppler and Brookfield viscosimeters. The electrical conductivity of polymer solutions was calculated by the equation $\lambda = \alpha/R$, Ohm^{-1} cm^{-1}; the electrical resistance was measured with E7–15 instrument.

The fiber diameter distribution was studied by microscopy (optical microscope, Hitachi TM-1000 scanning electron microscope). Fiber orientation was studied by using birefringence and polarization IR-spectroscopy (SPECORD M 80 IR-spectrometer). Crystalline phase of polymer was studied by differential scanning calorimetry (DSC) (differential scanning calorimeter). The packing density of fibrous materials was calculated as a function of airflow resistance variation with a special manometric pressure unit [7].

Physical-mechanical characteristics of fibrous materials were determined with PM-3–1 tensile testing machine according to TU 25.061065–72. Kinetics of UV-aging was studied with Feutron 1001 environmental test chamber (Germany). Irradiation of samples was carried out with a 375 W high pressure Hg-lamp, at a distance of 30 cm.

For in vitro biodegradation studies, the materials were incubated in test tubes filled with 10 mL of 0.025 M phosphate buffer solution (pH = 7.4) at 70°C for 21 days. At regular time intervals the materials were removed from the buffer and rinsed with distilled water, then placed into incubator at 70°C for 3 h, and finally weighed within 0.001 g.

15.3 RESULTS AND DISCUSSION

Diffraction patterns of the nanosized anatase and η-TiO$_2$ showed that sample S30 referring to anatase contains trace amounts of β-TiO$_2$ (JCPDS 46–1238). Analysis of the obtained size values of coherent scattering regions (*L-values*) of the η-TiO$_2$ and anatase samples showed that $L = 50$ (2) Å and $L = 100$ (5) Å, that is, crystallite sizes for the η-TiO$_2$ samples are substantially less.

Examination of the electrical conductivity of polymer solutions and choosing solvent mixture allowed the ($[CH_3(CH_2)_3]_4N$) – (TBAI) concentration in solution to be decreased from 5 down to 1 g/L due to adding the S1. The increase in PHB concentration up to 7 wt.% in a new solvent mixture was gained.

The fiber diameter distribution was studied by using fibers from 5% PHB solution in chloroform/formic acid mixture and 7% PHB solution as well. It was found that 550–750 nm diameter fibers are produced from 5% solution, while 850–1250 nm diameter fibers are produced from 7% solution, that is, the fiber diameter increases as the solution concentration rises (Fig. 15.1). The increase in the process velocity leads to the fiber diameter virtually unchanged.

FIGURE 15.1 Microscopic images of fibrous material as a function of PHB solution concentration: (a) 5%, (b) 7%.

Examination of fibrous materials by electron microscopy showed that the fibrous material uncovered by the current-conducting layer decomposes in 1–2 min, when exposed by electron beam. Examination of polymer orientation in fibers by using birefringence showed that elementary fibers in non-woven fabric are well oriented along fiber direction. However, it is impossible to determine the degree of orientation quantitatively because the fibers are packed randomly.

Analysis of fibrous materials by differential scanning calorimetry showed that at low scanning rates a small endothermic peak appears at 190÷200°C which indicates the presence of maximum straightened polymer chains (orientation takes place). Upon polymer remelting this peak disappears.

IR-examination showed that this method can be also applied only for qualitative estimation of orientation occurrence.

Measurement of the packing density of fibrous materials showed that the formulation used a day after preparing the solution has the greatest density of fiber packing. Probably, it is concerned with a great fiber diameter spread when smaller fibers are spread between bigger ones. The TiO_2-containing fibers are also characterized by the increased packing density.

The results of physical-mechanical testing of fibrous materials of different formulations are presented in Table 15.1.

TABLE 15.1 Physical-Mechanical Characteristics of Fibrous Materials Prepared from Polymer Compositions as a Function of Formulation

Formulation	Breaking length (L), m	Breaking elongation (ε), %
PHB + TBAI	724.43	20.83
PHB + TBAI (after a day)	1001.78	12.48
PHB + TBAI + S-12	1430.45	53.37
PHB + TBAI + S-30	1235.91	62.02

Physical-mechanical tests showed that introducing nanosized TiO_2 into solution substantially alters the properties of the resulting fibrous material.

Non-woven fibrous material contains randomly packed fiber layers. Actually, its deformation can be considered as a "creeping", and more and more fibers while straightening during deformation contribute to the breaking stress growing up to its maximum. In other words, the two processes take place simultaneously, namely fiber straightening and deformation of the fibers straightened.

Obviously, as the TiO_2 content increases the fiber flexibility rises due to the decreased crystallinity. This results in a greater amount of fibers simultaneously contributing to the breaking stress and, consequently, the greater slope of the curve.

Moreover, the addition of both S-12 and S-30 is likely to cause the formation of a firm fiber bonding at crossing points, possibly due to hydrogen bonding. Figuratively speaking, a network is formed. At the beginning, this network takes all the deformation stress. Finally, network breaking occurs followed by straightening and "creeping" of fibers, which does not affect the stress growth.

Examination of the crystallinity of fibrous materials, films, and PHB powder showed that main crystallite modification of both PHB powder and fibers and films melts at 175–177°C, that is, the morphology of crystals remains unchanged (Table 15.2).

For fibers, the low-melting shoulder at 160–163°C appears which confirms either the presence of smaller crystallites or their imperfection. For the S-12 based formulation, the low-melting peak disappears which is the evidence of more uniform distribution of the additive within the material. The broadened crystallization peak is observed for the S-12 based samples. The increased friction and decreased chain mobility lead to the obstructed crystallization and the decreased crystallization rate. The narrowed crystallization peak is observed for the S-30 based samples. Crystallization proceeds only where no contact with this additive is. Hence, the S-12 and S-30 have different energies of intermolecular interaction of PHB chains and the additive surface.

The crystallinity was calculated by the following equation:

$$\alpha_{\text{кр}} \frac{H_{n\text{л}}}{146} ,$$

where, $H_{пл}$ is melting heat calculated from melting peak area, J/g; 146 is melting heat of monocrystal, J [7]

The most loosely packed structure, i.e. the most imperfect, is observed for the formulation prepared a day after.

The S-12 based samples demonstrate excessive fiber bonding; the TiO_2 particles themselves obstruct crystallization. In the case of the S-12, the distribution is good. The S-30 obstructs crystallization to a lesser extent; the fibers crystallize worse, and the chains are not extended. Crystallization in fibers occurs upon orientation. The additive is not considered as a nucleating agent.

TABLE 15.2   Widths of Crystallization Peaks at the Scanning Rate of 20°C/min

Material	Heating, No.	Tm, °C	Crystallization peak width (at the onset), °C
PHB (starting)	1	174.9	18.61
	2	173.97	18.02
PHB fibers 1 g/L TBAI	1	176.83	20.9
	2	169.27–159.74	20.93
PHB fibers 1 g/L TBAI	1	176.94	26.24
a day after	2	171.98–163.14	22.8
PHB fibers 1 g/L TBAI	1	177.49	34.11
based on S-12	2	170.33	27.38
PHB film	1	177.21	18.27
from chloroform solution	2	171.95–163.1	18.57

UV-aging tests showed that TiO_2-modified fibers demonstrate greater UV-aging resistance. Although the induction period for S-30 based samples is less than that for other formulations, the UV-degradation rate for the S-30 based sample is comparable to that for the S-12 based one. For TiO_2-containing samples the increased thermal degradation heat after UV-aging is characteristic since the UV-treated TiO_2 acts as the initiator of both thermal and thermooxidation degradation due to OH-groups traveling to powder granule surface. The onset temperature of both thermal and thermooxidation degradation decreases due to UV-degradation proceeded not only within amorphous phase but also within crystalline phase. The TiO_2 acts as the initiator in UV-aging processes.

15.4 CONCLUSIONS

- Physical-mechanical characteristics of fibrous materials increase as the TiO_2 is introduced.
- The morphology of main PHB crystallites in powder and fibers is kept unchanged. However, the fibers show the low-melting shoulder (small and imperfect crystals). The TiO_2 obstructs crystallization.
- PHB fibers are characterized by strongly pronounced molecular anisotropy.
- S-12 based samples show the best thermal- and thermooxidation degradation stability as well as UV-aging resistance.
- The results obtained can be considered as the ground for designing new biocompatible materials, such as self-sterilizing packing material for medical tools or a support for cell growth, etc.

KEYWORDS

- **Electrospinning**
- **Nanosized titanium dioxide**
- **Polyhydroxybutyrate**
- **Polymer nanofibers**

REFERENCES

1. Dadachov, M. U.S. *Patent Application Publication.* US 2006/0171877.
2. Dadachov, M. U.S. *Patent Application Publication.* US 2006/0144793.
3. Fomin, V. A., & Guzeev, V. V. (2001). *Plasticheskie Massy, 2,* 42–46.
4. Bonartsev, A. P. et al. (2008). *J. of Balkan Tribological Association, 14,* 359–395.
5. Kuzmicheva, G. M., Savinkina, E. V., Obolenskaya, L. N., Belogorokhova, L., I. Mavrin, B. N., Chernobrovkin, M. G., & Belogorokhov, A. I. (2010). *Kristallografiya, 55(5),* 913–918.
6. Yu N, Filatov. Electrospinning of fibrous materials (ES-process) Moscow: *Khimiya,* 2001–231 p.
7. Barham, P. J., Keller, A., Otum, E. L., & Holms, P. A. (1984). *J. Master Sci. 19(27),* 81–279.

CHAPTER 16

A COMPREHENSIVE REVIEW ON APPLICATION, PROPERTIES AND STABILITIES OF CNT AND THE CNT SPONGES

SHIMA MAGHSOODLOU and AREZO AFZALI

CONTENTS

ABSTRACT

Wherever oil is produced, transported, refined, or released from natural seeps, there will be oil slicks. While most of these spills will be small, a few will be large enough to cause serious impact to the environment. The topics develop to describe the fate and behavior of spilled oil on water. In this chapter, the data sources for oil properties and the most important methods of cleaning it up are reviewed. The standard formulas used to describe these processes and the changing physical properties of the slick are discussed, along with possible variations in these formulas. After discussion about different methods of oil spill cleanup, various efficient sorbent which can be used for this purpose are reviewed. Then novel sorbent carbon nanotube sponges are investigated in detail.

16.1   INTRODUCTION

An oil spill is the release of a liquid petroleum hydrocarbon into the environment, especially marine areas, due to human activity, and is a form of pollution. The term is usually applied to marine oil spills, where oil is released into the ocean or coastal waters, but spills may also occur on land. Oil spills may be due to releases of crude oil from tankers, offshore platforms, drilling rigs and wells, as well as spills of refined petroleum products (e.g., gasoline, diesel) and their by-products, heavier fuels used by large ships such as bunker fuel, or the spill of any oily refuse or waste oil. Spilt oil penetrates into the structure of the plumage of birds and the fur of mammals, reducing its insulating ability, and making them more vulnerable to temperature fluctuations and much less buoyant in the water. Cleanup and recovery from an oil spill is difficult and depends upon many factors, including the type of oil spilled, the temperature of the water (affecting evaporation and biodegradation), and the types of shorelines and beaches involved. Spills may take weeks, months or even years to clean up.

Crude oil and refined fuel spills from tanker ship accidents have damaged natural ecosystems in Alaska, the Gulf of Mexico, the Galapagos Islands, France and many other places. The quantity of oil spilled during accidents has ranged from a few hundred tons to several hundred thousand tons (Deep water Horizon Oil Spill, Atlantic Empress, Amoco Cadiz) but is a limited barometer of damage or impact. Smaller spills have already proven to have a great impact on ecosystems, such as the Exxon Valdez oil spill because of the remoteness of the site or the difficulty of an emergency environmental response. Oil spills at sea are generally much more damaging than those on land, since they can spread for hundreds of nautical miles in a thin oil slick which can cover beaches with a thin coating of oil. This can kill sea birds, mammals, shellfish and other organisms it coats. Oil spills on land are more readily containable if a makeshift earth dam can be rapidly bulldozed around the spill site before most of the oil escapes, and land animals can avoid the oil more easily.

For example, the attention of the world was drawn to the recent unprecedented oil spill in the Gulf of Mexico. On the 20th of April, 2010, the British Petroleum (BP) Deep water Horizon oil rig in the Gulf of Mexico blew up, killing 11 workers and injuring 17 others. The spill lasted for about three months, released nearly 5 million barrels of crude oil to the Gulf of Mexico which then affected and killed huge populations of marine animals and after 8 months after the incidence, soiled 320 miles (510 km) of beaches and shorelines and after additional 8 months (one and a half year later), a total of 491 miles (790 km) of shorelines were affected. The oil industry has recorded many of such huge spills in the past: the wrecking of the Torrey Canyon in 1967, the Santa Barbara channel platform blowout in 1969, the Gulf of Mexico drilling rig incidents in 1970 and 1971, the grounding of supertanker Amoco Cadiz in 1978, the disaster of the Piper Alpha platform in the North Sea, and operation Desert Storm that caused the release of a huge quantity of oil into the Arabian Gulf in 1991. Others are the 1989 Exxon Valdez spill in Alaska, the 1999 Erika spill in France, the Prestige in Spain, 2002, and most recently the 2010 BP rig blowout in the Gulf of Mexico the world worst oil spill on marine water ever (Fig. 16.1) [1].

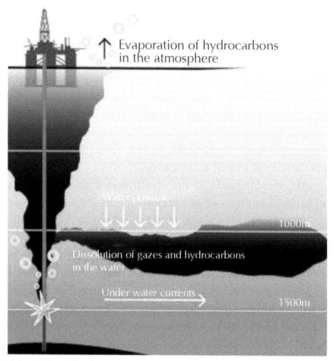

FIGURE 16.1   Oil spilling in water.

16.2 EFFECTS OF OIL POLLUTION

Oil penetrates into the structure of the plumage of birds and the fur of mammals, reducing its insulating ability, and making them more vulnerable to temperature alteration and much less floating in the water.

Animals that rely on scent to find their babies or mothers fade away due to the strong scent of the oil. This causes a baby to be rejected and abandoned, leaving the babies to starve and eventually die. Oil can mutilate a bird's fly ability, preventing it from foraging or escaping from predators. As they preen, birds may ingest the oil coating their feathers, irritating the digestive tract, altering liver function, and causing kidney damage. Together with their diminished foraging capacity, this can rapidly result in dehydration and metabolic imbalance. Some birds exposed to petroleum also experience changes in their hormonal balance, including changes in their luteinizing protein. The majority of birds affected by oil spills die without human intervention. Some studies have suggested that less than one percent of oil-soaked birds survive, even after cleaning, although the survival rate can also exceed 90 percent, as in the case of the Treasure oil spill.

Also, heavily furred marine mammals exposed to oil spills are affected in similar ways. Oil coats the fur of sea otters and seals, reducing its insulating effect, and leading to fluctuations in body temperature and hypothermia. Oil can also blind an animal, leaving it defenseless. The ingestion of oil causes dehydration and impairs the digestive process. Animals can be poisoned, and may die from oil entering the lungs or liver.

There are three kinds of oil-consuming bacteria. Sulfate-reducing bacteria (SRB) and acid-producing bacteria are anaerobic, while general aerobic bacteria (GAB) are aerobic. These bacteria occur naturally and will act to remove oil from an ecosystem, and their biomass will tend to replace other populations in the food chain.

Cleanup and recovery from an oil spill is difficult and depends upon many factors, including the type of oil spilled, the temperature of the water (affecting evaporation and biodegradation), and the types of shorelines and beaches involved.

The serious adverse effects of crude oil spills on economy, health and environment are much and regulatory measures put in place by government and relevant authorities in handling of crude oil only minimize the chances of these spills but do not eliminate it. As such the Nigerian petroleum industry expends so much resource on importing crude oil spill adsorbents for combating and cleaning of these minor spills. Estimated costs of spill clean up stand at 2 billion dollars per annum) [2].

In recent years, accidental and intentional oil discharges having occurred frequently during transportation, production, and fining, which results in severe negative impact on organisms and ecological environment [3]. Oil and chemical spill accidents can be caused by human mistakes and carelessness, deliberate acts such as vandalism, war and illegal dumping, or by natural disasters such as hurricanes or earthquakes. Offshore and shoreline waters can be polluted by oil drilling operations, accidents involving oil tankers, runoffs from offshore oil explorations and productions, and spills from tanker loading and unloading operations. Massive marine oil spills have occurred frequently and resulted in a great deal of damage to the marine, coastal and terrestrial habitats, economical impacts on fisheries, mariculture and tourism, and loss of energy source. Inland water bodies can be polluted by leaking of oil through pipelines, refineries, and storage facilities, runoff from oil fields and refinery areas and, in some cases, process effluent from petroleum refineries and petrochemical plants (Table 16.1) [4].

TABLE 16.1 Some Major Oil Spills and Corresponding Effects and Cleanup Techniques

Year	Place	Amount of oil spill (tones)	Affective cleanup techniques
1967	English Channel	120,000	High-explosive bombs were dropped on the wreck in an effort to start fires that would consume the remaining oil before it could spread.
			The use of straw and gorse to soak up the oil were used on many of the sandy beaches
1970	Nova Scotia	16,000	Absorbents — peat moss proved to be a good absorbent; straw was used on some beaches.
			In-Burning-used successfully on beaches and on isolated slicks in 10 to 2 °C water; part of spill was burned by spilling two drums of fresh oil and igniting
1978	France	230,000	On the shore, oil was removed mechanically and manually (most successfully with vacuum trucks and agricultural vacuum units).
			some of the rocky shore habitats were cleaned by pressure- washing with hot water
			Beaches were sprayed with artificial fertilizers and bacterial cultures.
			Rubber powder and chalk sinking agents were also applied although by this time the oil was too viscous and the powder could not mix with the oil.

TABLE 16.1   *(Continued)*

Year	Place	Amount of oil spill (tones)	Affective cleanup techniques
1989	Alaska	35,300	In-situ burning was successful but could not continue because of change in oil state as a result of the storm
			Sorbents were used where mechanical means were less practical because of costly logistics of skimmers far offshore. However, sorbents were labor intensive and generated additional solid waste.
			Warm water flushing of the beach used but the consequences were not favorable.
			Bioremediation enhancement agents were very effective in cleaning over 70 miles of shorelines.
1996	Sea Empress	70,000	50% of oil dispersed naturally
			some oil removed mechanically at sea
			use of chemical dispersant
2010	Gulf of Mexico	666,400	Booms and Skimmers.
			Dispersants.
			Controlled burning.
2011	Niger-Delta South Eastern Nigeria	1100	Soil sorption of crude oil seems to have profound effects on the soil properties analyzed for this study. Uzoije and Agunwamba reported that soil properties such as the bulk density, organic matter content and porosity were appreciably influenced.
2012	Louisiana coastline	2300	Many techniques were examined because of the critical situation
2013	Philippines, Estancia, Iloilo	1200	Many techniques were examined because of the critical situation

16.3 OIL PROPERTIES

While slicks of pure chemicals will not usually change their properties during the lifetime of the slick, this is not true for spills of crude oil or refined petroleum products. Because they are mixtures of hundreds of different organic compounds, each with its own unique characteristics, the nature and behavior of oil slicks evolves over time. This has important consequences with regard to the toxicity of the oil, its impact on the environment, and the effectiveness of different cleanup techniques [5].

16.3.1 DENSITY

The actual density of the oil will change over time. The lighter components will evaporate and the oil may emulsify, that is, water droplets become stabilized inside the oil, fanning an emulsion. Both of these processes will lend to increase the density of the oil and make it less buoyant. If the water into which it is spilled is cooler than the spilled oil, the oil density will further increase as the oil cools to water temperature. Nevertheless, under normal weathering processes the oil will still float unless sediment in the water column is mixed into it. This frequently occurs in the near-shore region, with the result that tannats can be seen on the bottom in the surf zone. As the slick ages, it often turns into tarballs. These tarballs arc only slightly buoyant and are subject to overwash by breaking waves. To an on-scene observer, it can appear that the oil is sinking. However, if the sea surface later turns calm, the tarballs will quickly resurface. Figure 16.2 shows a typical life of an oil slick.

Most weathering models use the approach of Mackay et al. [6], modified to account for emulsion formation [7], to forecast the change in oil density

$$\rho_e = Y\rho_w + (1-Y)\rho_{ref}\left[1 - C_1\left(T - T_{ref}\right)\left(1 + C_2 f_{evap}\right)\right] \qquad (1)$$

Here, C_1 and C_2 are empirically fitted constants that will vary somewhat for different oils. Reasonable values are 0.008 K^{-1} and 0.18, respectively.

Contiguous slick

Streaks and streamers

Tar balls

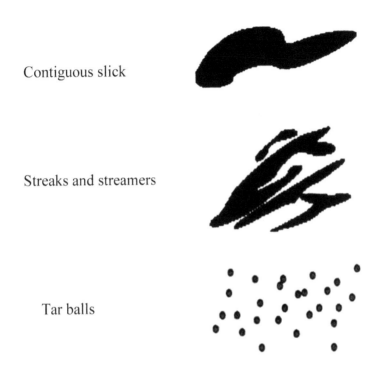

FIGURE 16.2   Evolution of an oil slick over time.

16.3.2 POUR POINT

Another field in many oil property databases is the pour point of the oil. This is defined as the lowest temperature at which oil will flow under specified conditions. Pour point is a difficult quantity to quantify and measurements of pour point vary widely.

While vaguely equivalent to melting point for pure substances, the pour point for oil, unlike the melting point of a pure chemical, will increase as the oil weathers. The most commonly used formula to describe this change is an algorithm proposed by Mackay et al. [6]

$$P = P_0 \left(1 + C_3 f_{evap}\right) \tag{2}$$

Here, C_3 is an empirically determined constant. Mackay et al. [7] used the value 0.35 for Prudhoe Bay crude.

16.3.3 VISCOSITY

Somewhat related to pour point is the viscosity of the oil, which is a measure of its resistance to flow. There are actually two closely related physical properties that bear the name viscosity: the kinematic viscosity, which has units of length squared, divided by time, and the dynamic viscosity which is the kinematic viscosity multiplied by the density.

Typically, once emulsified oil reaches such a high viscosity, its shear stress is no longer a linear function of the fluid velocity gradient, meaning that the oil has become non-Newtonian, acting more like a plastic semi-solid than a liquid. Viscosity measurement becomes dependent upon the shear rate. Cleanup of such emulsions presents considerable difficulties, since many pumps and skimmers are not designed to handle such thick fluids. Even before the oil has weathered to this state, it may be 100 viscous to be dispersible with chemical treatment.

Viscosity is a strong function of temperature. Therefore, it is necessary when using a historical value for oil's viscosity to know the reference temperature at which the viscosity was measured. A commonly used laboratory reference temperature is 100 °F, which is not a common temperature found at oil spills. Maekay et al. [6] recommend an exponential form for the temperature correction function:

$$V_0 = V_{ref} \exp\left[C_{VT}\left(\frac{1}{T} - \frac{1}{T_{ref}} \right) \right] \qquad (3)$$

As oil evaporates, the relative concentrations of the various components change.

16.3.4 SURFACE TENSION

Some spreading and dispersion algorithms require knowledge of oil's surface tension. Surface tension is the force of attraction between the surface molecules of a liquid. Chemicals that reduce surface tension can be used to facilitate dispersion. Laboratory data exist for the interfacial surface tension between oil and water and oil and air.

$$ST_w = ST_{w0} \left(1 + f_{evap}\right) \tag{4}$$

$$ST_A = ST_{A0} \left(1 + f_{evap}\right) \tag{5}$$

16.3.5 FLASH POINT

An occasionally important temperature for oils is the flash point, which is a measure of the flammability of the oil. Noting that the vapor pressure of each component is a function of temperature, the flash point is reached when

$$\sum_i x_i.MW_i.P_{vi}\left(T\right) = 104.7 \tag{6}$$

Here, the vapor pressure and molecular weight are expressed in MKS units and the sum is over the number of pseudo-components. This corresponds to requiring the sum of the partial pressure of the oil volatiles to be about 1% of an atmosphere. As the oil evaporates, the mole fractions of the less volatile components increase while those of the more volatile constituents decrease. In order to preserve the equality, the vapor pressures must be referenced to an increasing flash point temperature.

16.4 BEHAVIOR OF OIL SPILL PHENOMENA

After oil entering the water area, it will extend and form a big area oil slick under the role of gravity, inertia force, viscous and surface tension. Crude oil contains plenty of different ingredients. And every ingredient has different steam pressure, dissolvability, viscosity and surface tension. The group of high volatility evaporates very quickly, while the group of light carbon dissolves gradually in water, and the group that does not easily volatilize is left on the sea (Fig. 16.3). [8].

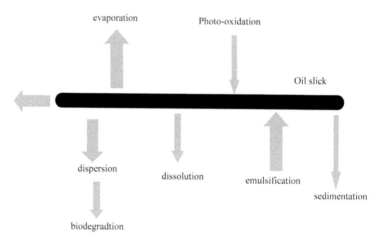

FIGURE 16.3   The major weathering and natural removal processes that can affect spilled oil on water.

When oil slick extends on the sea, wind, wave, tide and current make it drift. The dominant early behavior of oil slick on sea includes drift, evaporation, extend and dissolution. Few days after oil spill, dissolution and evaporation will make the oil volume decrease to almost half, especially evaporation. Along with the oil volume changes, ingredients of oil changes correspondingly. Hence, oil's viscosity and surface tension change consequentially. And all of this will influence the oil slick's behavior on the sea. Emulsification has an important influence on oil slick behavior. Water-in-oil is that water scatters with the form of particles in continuous oil slick. It makes the effective volume of oil slick large, changes oil's viscosity and makes oil become non-Newtonian fluid. There are some relationship between emulsification formation speed and the oil composition, ocean condition and temperature. Part of the oil particles may adhere to the particles floating in the rushing water and then sink to the bottom gradually. The remaining oil slick floating out of the surface may split into scraps while they are drifting and expanding, forming into coal tar puddle and then diminishes after sensitization oxidation, sediment, dissolution and biological degradation.

16.4.1 OIL SLICK EVAPORATION

Since the time oil spill starts, the oil slicks experience extending, spreading, drifting, coming to land, decompounding and finally attenuating. This process goes with the oil slick evaporation phenomenon, whose calculation equation is [8]:

$$F_v = \dfrac{\ln\left[1 + B\dfrac{T_G}{T}\theta\exp\left(A - B\dfrac{T_0}{T}\right)\right]}{\dfrac{T}{BT_G}} \tag{7}$$

A and B are nondimensional constants, and for representative crude oil, A adopts 6.3 and B adopts 10.3.

16.4.2 OIL SLICK EXTENDING

During the extending phase, in the initial stages of oil spill, spread occupied the dominant status, and evaporation had some influence on it. As time passed, dispersion effect of surface slick was bigger and bigger and so became the main form of oil slick extending movement. The oil extending is the synthetical effect of spread and dispersion. Point oil spill spread can be divided into following situations:

16.4.2.1 STATIC POINT CONTINUOUS SPILL SPREAD

If we suppose the spill flux of time, then the spill volume when time is smaller than spill end time becomes:

$$V_t = \sum_{i=1}^{n} Q_i \, \Delta t \left[1 - k(t - i\Delta t)\right] \tag{8}$$

When time is bigger than spill end time, the spill volume is:

$$V_t = V_j \left[1 - k(t - t_j)\right] \tag{9}$$

The spread dimension can be divided into three stages ($\Delta = 1 - \dfrac{\rho_0}{\rho_w}$):

Gravity spread stage:

$$l_{\mathrm{I}} = K_{\mathrm{I}} \left(\Delta g V \right)^{\frac{1}{4}} t^{\frac{1}{2}} \tag{10}$$

Viscidity spread stage:

$$l_{\mathrm{II}} = K_{\mathrm{II}} \left[\left(1 - \Delta \right) \Delta g \right]^{\frac{1}{6}} v_w^{-\frac{1}{12}} V^{\frac{1}{3}} t^{\frac{1}{4}} \tag{11}$$

Surface tension spread stage:

$$l_{\mathrm{III}} = K_{\mathrm{III}} \delta^{\frac{1}{2}} \rho_w^{-\frac{1}{2}} v_w^{-\frac{1}{4}} t^{\frac{3}{4}} \tag{12}$$

16.4.2.2   STATIC POINT TRANSIENT SPILL SPREAD

In fact, transient spill is the special case of continuous spill when t_j is adopted with 0, the spread dimension of each stage becomes:

- Gravity spread stage:

$$l_1 = K_1 \left(\Delta g V^2 \right)^{\frac{1}{4}} \tag{13}$$

- Viscidity spread stage:

$$l_2 = 2K_2 \left(\frac{\Delta g V^2}{\sqrt{V_w}} \right)^{\frac{1}{6}} t^{\frac{1}{6}} \tag{14}$$

- Surface tension spread stage:

$$l_3 = 2K_3 \left(\frac{\delta}{\rho_w \sqrt{V_w}} \right)^{\frac{1}{2}} t^{\frac{3}{4}} \tag{15}$$

16.4.2.3   THE DIFFUSION OF SURFACE OIL SLICK

Considered the aeolotropy of ocean, the diffusion of oil slick has a main direction s and a subordinate direction n, and the calculation of diffusing dimension is as follow:

$$d_s = \omega \sigma_s \tag{16}$$

$$d_n = \omega \sigma_n \tag{17}$$

$$\sigma_s = \alpha_s t^{1.17} \quad and \quad \sigma_n = \alpha_n t^{1.17} \tag{18}$$

16.4.2.4   EXTENDING OF STATIC POINT OIL SPILL

In the process of oil spread and diffusion of simultaneous effect, the extending dimension of oil slick can be considered as the superposition of spread dimension and diffusing dimension.

$$D_s = k_{eq}(d_f + d_s) \tag{19}$$

$$D_n = k_{eq}(d_f + d_n) \tag{20}$$

16.4.3   EMULSIFICATION AND DISSOLUTION OF OIL SLICK

When drifting on the sea, influenced by the impact of wave and wind, particles of oil disperse towards water, and particles of water also disperse towards oil constantly, and soon they come into the oiliness emulsion [9]. Emulsification is the reverse of dispersion. Rather than oil droplets dispersing into the water column, water is entrained in the oil and causes significant changes in the volume, density, and, especially, viscosity of the slick. It is not uncommon for the viscosity of an emulsified oil to be two or three orders of magnitude larger than the viscosity of the fresh oil. This has important implications for cleanup policy, as many common cleanup tools may be rendered ineffective when the oil becomes too viscous (Fig. 16.4).

FIGURE 16.4   Two different possible water droplet distributions in emulsified oil.

Comparing to oil, oiliness emulsion is more viscid, and the volume is enlarged by 3–4 times, so that the pollution area is larger consequently. And because of the high viscosity, it is difficult for recovery work. For emulsification equation we adopt the formula given by Mackay:

$$\frac{dF_{wc}}{dt} = C_4 U_F^2 \left(1 - \frac{F_{wc}}{C_5} \right) \tag{21}$$

16.5   USE OF VARIOUS PROCESSES IN OIL SPILL RESPONSE TECHNIQUES

The various cleanup methods in practice include in-situ burning, mechanical methods (removal by physical means using skimmers [10, 11], vacuum units [12, 13], and booms [14–16]) chemical methods (use of chemical dispersants [17]), and sorbents (mineral products [18, 19], agricultural products [20, 21], and synthetic products [22]). There is no simple procedure, which can be recommended for all spills. Spilled oil will behave differently depending on the type of oil, the surface on which it spills, the soil and subsoil conditions, and the prevailing weather conditions. Hence, the choice of cleaning method must take into account these factors. In most cases, two or more methods are combined to achieve an effective cleanup [23]. The different cleanup techniques are discussed as follows:

16.5.1 OIL SORPTION

One of the most important methods for conflicting oil spills is oil sorption by sorbents, which is economical and efficient. Oil sorbents are able to concentrate and transform liquid oil to the semi solid or solid phase, which can then be removed from the water and handled in a convenient manner without significant oil draining out. There are some factors that make a sorbent material be suitable such as besides being inexpensive and readily available, demonstrate fast oil sorption rate, high oil sorption capacity (oleophilicity or lipophilicity), low water pickup, high oil retention capacity during transfer, high recovery of the absorbed oil with simple methods, good reusability, high buoyancy, and excellent physical and chemical resistances against deformation, photo degradation, and chemical attacks [4].

All these process influence the choice of oil-spill countermeasures. Quickly collection of the oil after a spillage and mechanical recovery by sorbents is one of the most important countermeasures in marine oil-spill response. Sorbents are solid products capable of trapping liquid pollutants. Sorbents are used to reduce the spread of a spill of pollutant, fix a pollutant by impregnation to facilitate its, recovery for small spills recover the pollutant from effluents generated, by cleanup operations filter pollutant that cannot be recovered from, a water mass (channels, rivers, water intakes and washing effluents) [24].

Two broad categories of sorption phenomena, adsorption and absorption, can be differentiated by the degree to which the sorbet molecule interacts with the sorbent phase and its freedom to migrate within the sorbent. In adsorption, solute accumulation is in general restrict to the surface or interface between the solution and adsorbent. In contrast, absorption is a process in which solute, transferred from one phase to the other, interpenetrates the sorbent phase by at least several nanometers. Sorption results from a variety of different types of attractive forces between solute, solvent and sorbent molecules. Chemical (covalent or hydrogen bonds), electrostatic (ion–ion, ion–dipole) and physical (Coulombic, Kiesom energy, Debye energy, London dispersion energy) forces act together, but usually one type prevails in a particular situation [25].

The addition of facilitate of sorbent materials in an oil spill area is a transformation from liquid phase to semi-solid phase. Once this change is achieved, the oil removal by the sorbent structure removal is facile. While hydrophobicity and oleophilicity are primary determinants of successful sorbents, other important parameters include retention over time, the recovery of oil from sorbents, the amount of oil sorbed per unit weight of sorbent, and the reusability and biodegradability of sorbent [18, 26, 27]. Therefore, it is essential to collect and clean the oil promptly after a spillage. Advanced removal and recovery of oil by oil sorbent materials is of great interest from economical and ecological standpoints and therefore, various materials have been applied to this end [28].

The sorbent acts as an obstacle to spreading and therefore constitutes an additional containment means. When in contact with a pollutant, the sorbent soaks up the pollutant like a sponge. The sorbent and pollutant mixture is then recovered. If the sorbent is far from saturated with pollutant, the pollutant has a good chance of being definitively fixed.[24]

16.5.2 IN-SITU BURNING OF OIL SLICK

Oil on water or between layers of ice can also be tackled quickly, efficiently and safely by controlled burning. This technique works most efficiently on thick oil layers, so the oil needs to be contained by pre resistant booms, ice or by a shoreline. On average, about 80–95% of the oil is eliminated as gas, 1–10% as soot and 1–10% remains as a residue. Following the burning, this residue can be recovered from the water surface. Controlled burning is a proven response developed over several decades by the oil industry, emergency response authorities and scientists. This involved extensive laboratory and tank testing, large-scale field experiments and lessons from real incidents.

In situ burning is the term used for a controlled burning of oil "in its original place" and refers to a technique in which accidentally spilled oil is ignited and burned directly on the water surface. The most amount of the oil is thereby removed from the spill site by transmuting it into inflammation products, soot and residue. In general, the method is not very labor-intensive, can easily be achieved and is efficient. Efficiencies of 98%

and 99% can be obtained under ideal circumstances (fresh oil and a thick slick). After flame out, only small percentage of the original amount of oil is left, which consists of a high viscous/semi-solid residue. The first recorded in situ burning was done in 1958 in Northern Canada. Since then, a series of burn and oil spill studies have been conducted, with a total of approximately 11 larger in situ burnings (accidental and experimental) in conditions with ice. To secure a successful burning it has been found that two conditions are especially important: the thickness and weathering of the oil. Of all the weathering processes, water-in-oil emulsification is of particular importance, however, the amount of light components left in the oil and spreading is also important. In addition, the type of oil and ambient conditions are crucial and too much wind will negatively affect the burning [29].

Some of the drawbacks with in situ burning are smoke production and the risk of secondary fires during a burn, as well as a concern for the workers and possible settlements nearby. Nevertheless, in situ burning may be the only method that works in ice-covered waters. Since conventional response technologies (mechanical and chemical) could have a limited potential in snow, ice and cold, whereas in situ burning is strengthened under such conditions. In the 1970 s, field trials showed that in situ burning had the potential for removing large amounts of oil on the ice surface. Later experiments were performed with in situ burning in ice and snow, primarily as laboratory and mesoscale experiments (tank or field tests). Hence, the focus has been on burning physics (flame spreading, oil spreading, slick thickness, burning efficiencies) and the study of burning processes of emulsions, development of igniters and the influence of wind and waves in small-scale field and tank tests. Despite the research conducted over the past decades, the use of in situ burning as an operational response tool is still not fully developed. There is a need for large-scale field experiments to compare and verify the results from the smaller scales, and an evaluation of certain techniques and tactics that can only be done in an actual oil burning situation [29].

In-situ burning is effective if the following conditions are in place: (i) the oil slick is sufficiently wide so that a good volume of oils burnt off at a time, (ii) the oil is very thick to sustain combustion, (iii) the water is calm and (iv) the slick location is distant from sensitive facilities. However, because of sea weather conditions discussed above (sea current, wind and temperature), all of these conditions rarely exist for a long time and for this reason in-situ burning is limited by the following limitations. However, a strong advantage of in-situ combustion over conventional spill cleanup techniques is in ice or cold water application where mechanical booms and chemical dispersants have limited efficiency whereas all conditions favoring combustion tend to persist for a long time since ice strongly influences weathering. The more the ice concentration on sea surface, the less the weathering by evaporation, and the more the ices limit oil spreading keeping the oil slick thick enough for burning. This is true because atmospheric and sea condition in ice water is expected to be characterized by low or no tidal currents and low temperature all of which are in favor of combustion [1, 30, 31].

16.5.2.1   LIMITATIONS OF IN-SITU BURNING

Safety hazards for in-situ burning operations should be similar to those of ordinary skimming at sea. In-situ burning is normally done as early as possible, before evaporation and natural dispersion occur. In reality, there are a number of problem with the added hazards related to the combustion process:

In-situ burning is a process that involves the intentional setting of a fire. Great care must be taken so that this fire is controlled at all times.

1. Ignition of the oil slick, especially by aerial ignition methods (such as the helitorch), must be well coordinated with neighboring vessels and be carefully executed. Weather and water conditions should be kept in mind, and proper safety distances should be kept at all times.

2. In-situ burning at sea will involve several vessels working relatively close to each other, perhaps at night or in other poor-visibility conditions. Such conditions are hazardous by nature and require a great degree of practice, competence, and coordination.

3. Response personnel must receive the appropriate safety training. Training should include proper use of personal protective equipment, respirator training and fit testing, heat stress considerations, first aid, small boat safety, and any training required to better prepare them to perform their job safely.

Safety hazards are substantial and should be given due attention. In-situ burning will require only a fraction of the people needed to clean the same amount of oil if it impacts the beaches. In addition, personnel conducting the burn are expected to be well trained and monitored and, hopefully, have a low accident rate. In-situ burning, by minimizing the amounts of oil impacting the beaches, may prevent the illnesses and injuries that are often associated with beach cleanup operations [32].

Fresh crude oil has to be at least 1 mm thick before it can be ignited on water, whereas oil that has undergone extensive weathering may need to be at least 2–5 mm. Heavy fuel oils need to be contained to maintain at least a 10 mm or nearly one-half an inch slick thickness. Once ignited, a burn will continue until the oil slick is less than about 2–3 mm, or about 1/10 of an inch thick. The reason that slick thickness is so important for on-water burning is that very thin slicks are rapidly cooled by loss of heat to the under lying water. To burn, oil must be heated enough to vaporize some of the oil's components. It is actually the oil vapors forming above the slick that burn, not the liquid oil. A thicker oil slick acts as an insulating layer, whereas, a thin oil slick cools to the point that vapors are not formed and the fire goes out. Experiments have been conducted to measure the thickness at which the fire goes out, so these "rules of thumb" are generally well accepted [31, 33].

Complete removal may not be receivable because of the prevailing conditions in the sea: the cooling effect of wind and wave action, which may rise high and extinguish the fire even if booms are used to contain the slick [31].

Large amounts of smoke from oil slick burning can result in oil rain. The formation and possible sinking of extremely viscous and dense residues can damage the sea bed and its inhabitants. The viscous residue may also be transported to shorelines and beaches by ocean tides or currents. Airborne irritants and possibility of secondary fire are sources of concern when combustion has to be carried out close to residential areas. Carbon monoxide, sulfur dioxide, and polycyclic aromatic hydrocarbons (PAH) are common toxic compounds emitted while burning oil on water.

16.5.3 MECHANICAL TECHNIQUES

If a spill does occur, the rapid containment and recovery of oil at or near the source is the first goal. Mechanical skimmers can be used to remove oil from the water surface and transfer it to a storage vessel. Skimmers work most efficiently on thick oil slicks: floating barriers, known as oil booms, are used to collect and contain spilled oil into a thicker layer. A variety of designs for skimmer and booms have been adapted for sea conditions and some of the presented significant efficient.

Mechanical methods involve the use of booms spread over surface of seas, estuaries and coastal waters to prevent the spread of oil slicks or to direct their movements. Booms are combined to make "V" shaped barriers, which concentrate the oil for pickup by skimmer barges and boats. The advantage of booms and skimmers over other commonly used methods such as chemical dispersants and in-situ combustion is the absence of adverse environmental effects [1, 34].

Mechanical recovery is the most commonly used oil spill response technique. This technique physically removes oil from the water surface, even in the presence of ice [35]. Unlike other cleanup techniques, mechanical recovery can be efficiently applied to treat emulsified oils as well as oils of variable viscosities. A weakness of mechanical cleanup is the recovery rate. It may be very time consuming and expensive when employed on a large scale, and require a large amount of personnel and equipment, and every additional hour of cleanup time can significantly increase the cost of recovery. A more efficient recovery device can thus reduce the cost significantly and reduce the risk of oil reaching the shoreline [31, 36].

The adhesion (oleophilic) skimmer is one of the most common types of mechanical recovery equipment. It is based on the adhesion of oil to a rotating skimmer surface. The rotating surface lifts the oil out of the water to an oil removal device (e.g., scraper, roller, etc.). A number of studies have been undertaken to test the recovery rate of various skimmers [34, 37].

16.5.3.1   LIMITATIONS OF MECHANICAL TECHNIQUES

Oil spill cleanup by mechanical technique is expensive, and requires large number of personnel and equipment. Some mechanical limitations of booms generally include attrition under harsh sea conditions and escape of oil underneath the boom at slick velocity in excess of one knot. An important structural limitation, which occurs when the effective boom draft is lower than the oil slick thickness resulting in escape of some oil below the barrier, is boom drainage failure. Other structural limitations of booms, that can be mentioned here, are failure mechanisms: the droplet entrainment failure and critical accumulation failure. Booms and skimmers are also expensive to operate when they have to be deployed far offshore. Furthermore, poor efficiency results can result in higher cost of spill cleanup. In summary, booms are only effective in calm water conditions with little wind or currents such as littoral waters, estuaries and port basins. However, their structural designs also have a great impact on their performance efficiency.

16.5.4   BIOREMEDIATION

Biodegradation is a process by which small organisms like bacteria, yeasts, and fungi break up complex compounds into smaller compounds for their food. This process occurs naturally. Its application in oil spill cleanup involves the artificial introduction of biological agents such as fertilizers and nutrients to native microorganisms in the contaminated site so they proliferate (bio-stimulation) or the introduction of non-native microorganisms (bio-augmentation) to speed up the natural process of biodegradation so as to protect shorelines, wetlands, and other marshy areas affected by spills from further damage. This process is known as bioremediation. Proof of its effectiveness as an oil spill cleanup technology was developed on the shoreline of the Delaware Bay.

16.5.4.1   LIMITATIONS OF BIOREMEDIATION

Bioremediation is ineffective in removing oil spills that consist of large coherent masses or for sunken oil spills. Bioremediation is also limited by biotic environmental factors such as a low level of nutrients including phosphate and fixed forms of nitrogen, very low temperatures, and insufficient oxygen.

16.5.5   DISPERSANTS

Chemical dispersants are another method of cleaning up spills. They have proven highly effective in the Arctic through extensive testing. Dispersants are like detergents designed for use in marine environments. They accelerate the breakup of oil slicks into fine droplets that can then disperse and biodegrade more easily in the sea. The use of dispersants offshore is generally recognized as an efficient way of rapidly treating large areas of spilled oil to reduce the impact on marine life and the environment. They can be applied from fixed-wing aircraft, helicopters, and vessels.

Dispersants are able to treat larger areas compared with other methods but their use should be restricted to sufficiently deep water where proper agitation will result in rapid dilution in the upper column of the water body and the toxic effect will be minimal at the sea bed.

Dispersants consist of different surfactants (surface active, "soap-like" molecules). Surfactants are partially soluble in both oil and water. When sprayed on an oil slick, surfactants reduce the interfacial tension between the oil and water. This enhances dispersion and increases the natural dilution and biodegradation process of oil in water. Surfactants are generally applied by spray equipment followed by excitation to mix the chemical with the oil for maximum effectiveness. Wind therefore plays an important role in the mixing. The results of the effectiveness of dispersant spraying techniques and the need to understand the weathering process of an oil slick before spraying. It should be considered that the length of time dispersants effectively sprayed on an oil slick in calm water will to be effective. Their studies became important considering the low effectiveness of chemical dispersant in calm water where there is no sufficient energy to break the oil and water whose interfacial energy has been greatly reduced by dispersant. However, ecological considerations, experience, and technological developments in the handling of oil spills have pushed chemical dispersants very much out of the picture. For instance, dispersants have not been used extensively in the United States because of difficulties with application, disagreement among scientists about their effectiveness, and concerns about the toxicity of the dispersed mixtures and in the United Kingdom, the use of dispersants is a regulated activity. This could affect the fisheries food industry and, hence, have a health impact on people including respiratory, nervous system, liver, kidney, and blood disorders.

Allen and Ferek did a cost comparison using representative mechanical, dispersant, and burning systems for the recovery/elimination of approximately 8000–10,000 barrels (1272–1590 m^3) of oil in a 12-h period: mechanical, $100–$150 per barrel of oil, dispersants, $50–$100 per barrel, and in-situ burning, $20–$50 per barrel.

16.5.5.1   LIMITATIONS OF DISPERSANTS

Dispersants are expensive and contain toxic compounds harmful to aquatic fauna and flora. Furthermore, they are ineffective in calm water where is no sufficient mixing energy needed to mix dispersants with oil and to also aid immediate dispersion of the oil. They are also more effective in thicker oil slicks than thinner ones because the dispersants are easily lost in thinner slicks. Also, thicker slicks subjected to weathering action will become more viscous and thereby reduce the effectiveness of the dispersants, though the effect is less severe than dispersant loss in thick slick.

16.5.6   REMOVING BY SORBENTS

Sorbents are products or materials that are oleophilic and hydrophobic, they have a high capacity to sorb oil and repel water. There are three classes of sorbents synthetic organic, inorganic mineral and agricultural (organic) products. The sorbent material is broadcast over a slick and allowed to sorb oil. The oil-soaked material is then collected and, depending upon the sorbent, the sorbent will be squeezed to remove oil and then rebroadcasted, or the oil-laden material will be disposed of safely. The efficiency of a sorbent depends on its recyclability, wet ability, density, geometry, sorption capacity and sorption rate. These properties determine the time required to spread and harvest the sorbents. A common requirement for all sorbents is that they must be spread on the spill before the oil viscosity increases (due to evaporation of volatile components) to the point that sorption is no longer possible. The advantage of sorbents is their insensitivity to sea conditions. Sorbents have been recorded to be one of the most effective and cheapest methods of cleaning oil spills on shorelines whose contamination has always had the highest economic and environmental impact because of the difficulty in cleaning oil spilled on them.

16.5.6.1   INORGANIC MINERAL SORBENTS

Inorganic materials also known as sinking absorbents are highly dense, fine grained mineral materials either natural or processed used to sink floating oil. Examples include stearate-treated chalk and silicone-treated pulverized fly ash, zeolites, graphite, activated carbon, organoclay, silica (sand) and silica gel.

They are generally disliked as they have numerous shortcomings such as contamination of sea beds and harmful effects to aquatic habitats. They also tend to release some of the absorbed oil while sinking because of low retention capacity of some of the solids [30].

The sorbent materials in use for oil spill cleanup include inorganic natural such as inorganic oil sorbent materials CF_3-modified silica aerogels, pure-or high-silica synthetic zeolites and geopolymers, organophilic clays, exfoliated graphite, expanded perlite, activated carbon have already been investigated.

The use of inorganic mineral products as oil sorbents requires in most cases their chemical or surface modification, in order to ameliorate their hydrophobic character and affinity for organic compounds. For example, modification with CF_3–$(CH_2)_2$-groups can be used, in order to limit the structure collapse of silica aerogels due to water adsorption, resulting in an excellent oil sorption capacity. Cationic surfactants, such as quaternary ammonium cations can be applied in order to ameliorate the oil sorption capacity of inorganic materials, such as zeolites, clays and fly ash [30–38].

One of the most important sorbent because of its special properties, in this class, is activated carbon. For example, organoclays are hydrophobic and have high sorption and retention capacity but they are not degradable and cheap [39] but activated carbons (particularly those of agriculture origin) are cheap and readily available from many companies. They owe their distinguished properties to an extensive surface area, high degree of surface reactivity and favorable pore size distribution. It has good sorption capacity. However, granular organoclay can be seven times more effective than activated carbon. Hence, organoclays can be used to improve the sorption efficiency of activated carbon used activated carbon enhanced with organoclay to clean hydrocarbon spill in water. Activated carbon is widely used in oil spill cleanup. Many commercial sorbents incorporate activated carbon in the sorbent pads to facilitate cleanup. For example, one common type consists of two sheets of cotton with activated carbon sandwiched between them. The activated carbon separates and holds toxic parts of the oil such as the polycyclic aromatic hydrocarbons, protecting spill cleanup workers. Arbatan et al. (2011) studied the oil sorption capacity of calcium carbonate powder treated with fatty acid to change the wet ability of the carbonate powder from water wet to oil wet. Results showed that the treated calcium carbonate powder to be very hydrophobic and selectively absorbed diesel and crude oil out of oil water mixture. Although, Calcium carbonate is a natural material and not known to be harmful, however, the recoverability and re-usability of the calcium carbonate sorbent after its saturation with oil.

It is sometimes difficult to determine which class of sorbents do activated carbon belongs because it is of either botanical origin (wood, coconut shells, fruit seeds and stones) [40–42], mineral origin (coal, lignite, peat, petroleum coke) [43–45], or polymeric material origin (rubber tires, plastics) [46–48].

A study of exfoliated graphite indicates their high heavy oil sorption capacity compared to polypropylene mats, perlite, cotton, milkweed, and kenaf. Perlites have also proven to have a sorption capacity less than, but comparable to, most synthetic sorbents for oil spill cleanup.

16.5.6.1.1   LIMITATIONS OF MINERAL SORBENTS

Mineral sorbents are generally disliked as they have numerous shortcomings such as contamination of sea beds and harmful effects to aquatic habitats. They also tend to release some of the absorbed oil while sinking because of the low retention capacity of some of the solids. Other disadvantages of activated carbon include fire risk, pore clogging, and problems with regeneration. Another limitation of mineral sorbents (apart from those of agriculture origin) is that they are very expensive and are not commonly used.

16.5.6.2   SYNTHETIC ORGANIC PRODUCTS

The most widely used sorbents are synthetic sorbents made from high molecular weight polymers, such as polyurethane and polypropylene. They are available under various trade names. They have good hydrophobic and oleophilic properties and high sorption capacity. For example, ultra-light, open-cell polyurethane foams are capable of absorbing 100 times their weight in oil from oil-water mixtures. Also, the oil sorption efficiency of tire powders and its applicability in oil spill cleanup. Their study has been shown that tire is oleophilic and can absorb 2.2 g of oil per unit gram of the sorbent. Because of the re-usability of the tire sorbent e as much as 100 times without the tire powder losing its sorption capacity e tire powder is able to absorb as much as 220 g per gram of sorbent after 100 cycles of usage. However, tire is not biodegradable and thus its usage for oil spill control will be of environmental concerns and the cost of the mechanized system is an additional [30].

16.5.6.2.1   LIMITATIONS OF SYNTHETIC ORGANIC SORBENTS

The non-biodegradability of synthetic sorbents is a major disadvantage. As stated earlier, synthetic sorbents are not biodegradable but newer concepts enable polyurethane foam to be broadcast, recovered, cleaned, and reused in a totally mechanized process, thus removing the necessity for disposal and need for biodegradability. For example, one style consists of a floating rope of sorbent material that is freely deployed on the water surface. The rope is drawn through an oil slick, picking up the oil. It is brought aboard a support vessel, passed through squeeze rollers to remove the recovered oil, and then re-deposited on the water surface in a continuous operation. Nevertheless, the mechanized system is an additional cost.

16.5.6.3   NATURAL ORGANIC (AGRICULTURAL PRODUCTS)

The main advantages of natural organic sorbents involve their low-cost, availability as renewable and abundant resource supply in nature, good oil sorption performance and their property to be "ecofriendly" (assimilated by environment after usage, biodegradation). Organic natural materials involve many agricultural products, like cotton fiber [49], kapok fiber [3, 4], straw, rice hull [50] wood sugarcane bagasse[50–53], kenaf [54, 55], cotton, cotton grass fiber [54–56], corn cobs, sawdust [52], wool-based sorbents [57, 58], peat moss, milk weed [55], pine bark, banana pith corn stalk [52], water hyacinth roots, chitosan [59], bentonite [60], and activated carbon [2], recycled wool, silkworm cocoon wastes [21], felt, coconut shells and rice husks [21, 52], populus seed fibers [61], waste coir pith [52], various vegetable fibers (leaves residues, sawdust, sisal, coir fiber, sponge-gourd, silk-floss) [25]. It has also been shown that some agriculture products like straws, cellulosic fiber, milkweed [54, 55]and cotton fiber sorbents can remove significantly more oil than polypropylene (synthetic organic) materials used commercially.

The potential of raw law value cotton fibers, to remove used oil can be used to provide an efficient, easily deployable method of cleaning up oil spills and recovering of the oil. It is important to provide a safe system for oil removal and recovery. It is proven that the loose low grade cotton fibers and the pad have an excellent commercial potential as a sorbent for oil.

Among of those natural materials, kapok has the advantages over traditional oil-absorbing materials: low cost, biodegradability, intrinsic hydrophobic characteristic and high sorption capacity, and accordingly, they are preferable as an oil-absorbing material. Kapok, an agricultural product, is a fiber derived from the fruits of silk-cotton tree, and is mainly composed of cellulose, lignin, polysaccharide, besides these constituents, a small amount of waxy coating also covers the fiber surface makes it much hydrophobic. In recent years, kapok fiber has received increasing attention as an oil-absorptive material due to its distinct hollow structure and hydrophobic characteristics. In these studies, kapok fiber exhibited good water repellency, high oil adsorption capability, and well reusable characteristics, demonstrating its potential as an alternative for application in oil pollution control. However, oil absorption mechanism, the contribution of hollow lumen and surface wax on the oil absorption capability of kapok fiber still cannot be well recognized [3–4, 62–63].

Vegetable fibers are environmentally friendly materials, with densities close to that of synthetic polymers or even lower, and may show high oil sorption capacity at a usually low cost. The use of various vegetable fibers, namely mixed leaves residues, mixed sawdust, sisal (agave sisalana), coir fiber (cocos nucifera), sponge-gourd (luffa cylindrica) and silk-floss as sorbent materials of crude oil has an outstanding sorption rate and sorption capacity of 85x its weight. It also exhibited goo buoyancy, hydrophobicity, oleophilicity and also a good retention capability [25].

Synthetic material polyurethane foam is a much better sorbent than agriculture products and has proven to have the overall highest sorption capacity 100 times its weight. However, the natural materials mentioned above (such as straw, Kapok, vegetables fibers, peat, or bark) are more readily available at much lower costs and biodegradable.

16.5.6.3.1 *LIMITATIONS OF NATURAL ORGANIC (AGRICULTURE) SORBENTS*

The limitations of agricultural sorbents include high cost. This is due to the cost involved in recovering the oil soaked sorbent, removing the oil, and re-dispensing the sorbent. For example, a million gallons of oil spilled will require 200 tones (65,000 bales of straw) to sorb the oil. This means that after use and harvesting (recovery), vessels on location must be able to store 20 times the original weight of the sorbent since the sorbents are now soaked with oil and water. An ocean-going vessel with large amounts of deck space for ad hoc storage and sundry material handling equipment that could shuttle back and forth between land and a mother ship without losing any response effectiveness will serve the purpose.

Furthermore, straw and other agricultural products require spreading of dry sorbents and retrieval of soaked ones by hand labor, which is time consuming and costly. Hence, their application is limited to small terrestrial or marine spills or cleanup of residual spills after major clean up operations by other techniques like in-situ burning.

A sorbent is considered reusable (recyclable) if a loaded sorbent can be easily compressed or squeezed to its original size and shape. Other limitations of agriculture sorbents are their relatively lower sorption capacity (compared to polyurethane and exfoliated graphite) and also their limited recyclability. However, as discussed previously, some natural sorbents such as Kapok, silk-floss fiber, straw sorb significantly more oil than some synthetic materials used commercially, like polypropylene.

16.5.6.4   CARBON NANOTUBES SPONGE (CNT)

Carbon nanotubes (CNTs) have a unique one-dimensional structure, large specific surface area, and are oleophilic and hydrophobic [64]. As filter materials, CNTs exhibit selective separation characteristics for different solutions [65, 66]. The CNTs directly synthesized on the surface of expanded vermiculite could significantly improve the oil sorption capacity by forming a hydrophobic surface [67].

Here we will investigate a highly porous CNT material (CNT sponges) as sorbents for oil absorption. For this purpose, CNTs will be discussed, first. After that CNT sponges will be reviewed.

16.5.6.4.1   CNTS: STRUCTURE, PROPERTIES

Carbon is a really versatile element, for reasons rooted from its four valence electrons and the right atomic size. It can form numerous compounds, many of which are the basis of life on this planet. Even pure carbon can have quite a few allotropes. This is because the four valence electrons can make different types of bonds with other carbon atoms. Diamond and graphite (upper left and right in Fig. 16.5) have distinct optical, electrical and monetary properties, all just because the carbon atoms arrange themselves in different ways. It was not until about 20 years ago before people gained the ability to probe structure of nanometer scales and more interesting forms of carbon were discovered.

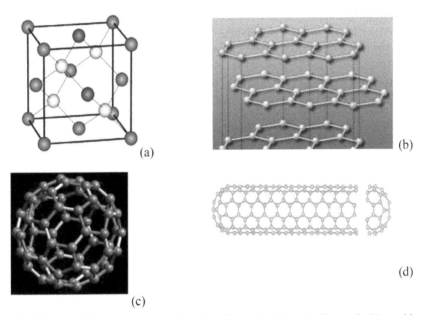

(a)

(b)

(c)

(d)

FIGURE 16.5   Models of carbon allotropies. Clockwise from (a) diamond, (b) graphite (c) C60 buckyball, (d) nanotube.

The zero dimensional C60 buckyball was discovered in spectroscopy data in 1985 [68], followed by one dimensional nanotube in 1991 [69]. These newly-found carbon structures couple quantum effects, lower dimensionality and the unique properties of graphene [70–71] all together, and they have generated intense research in many disciplines because they bridge between physics, chemistry, material science and more.

CNTs can be multi-walled with a central tubule of nanometer diameter surrounded by graphitic layers separated by ~0.34 nm. Carbon nanotubes are hollow cylinders with diameters ranging from 1 nm to 50 nm and length over 10 μm. They consist of only carbon atoms and can be thought of as a graphite sheet that has been rolled into a seamless cylinder.

By contrast, single-walled CNTs (SWCNTs) are a cylindrical tube formed by wrapping a single-layer graphene sheet. According to the wall number of a CNT, it can be classified as a SWCNT, double-walled CNT (DWCNT), triple-walled CNT (TWCNT), or multi-walled CNT (MW-CNT).

SWCNTs can be considered as rectangular strips of hexagonal graphite monolayers rolling up to cylinder tubes. Two types of SWCNTs with high symmetry are normally selected by researchers, which are zigzag SW-CNTs and armchair SWCNTs. When some of the atomic bonds are parallel to the tube axis, the CNT is called a zigzag CNT, while if the bonds are perpendicular to the axis, it is called an armchair CNT, and for any other structures, they are called chiral CNTs, as shown in Fig. 16.6 [72].

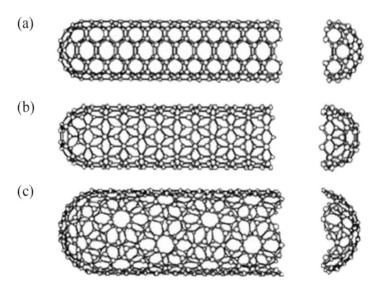

FIGURE 16.6　Some SWCNTs with different chiralities. (a) armchair structure (b) zigzag structure (c) chiral structure.

Because of the covalent sp² bonds between individual carbon atoms, a nanotube can have a Young's modulus between 1.2–5.5 TPa, a tensile strength about a hundred times greater than that of steel, and can tolerate large strains before mechanical failure. A CNT can be a metal, semiconductor, or small-gap semiconductor. The state of the CNT depends heavily on the (m, n) indices, and, therefore, on the diameter and chirality. CNTs also possess tunable surfaces characteristics, well-defined hollow interiors, and high biocompatibility with living systems. Based on these properties, many potential applications, including both large-volume applications (such as conductive, electromagnetic, microwave absorbing, high-strength composites, super capacitor, or battery electrodes, catalyst and catalyst support, field emission displays, and transparent conducting films) and limited-volume applications (such as scanning probe tips, drug delivery systems, electronic devices; sensors; and actuators) have been proposed [73]. Controllable mass production of CNTs with the desired structures and properties is essential for future applications. In the past 20 years, arc discharge, laser ablation, and chemical vapor deposition (CVD) methods have been developed to produce CNTs in sizeable quantities. In CVD methods, the catalytic decomposition of a carbon feedstock into carbon and hydrogen atoms is initiated on an active catalyst surface, where the tubular CNTs grown. CVD growth can be achieved under mild conditions (such as normal pressure and low growth temperature) at a low cost. The wall number, diameter, length, and alignment of these CNTs can be well mediated. Thus, the CVD method is the most promising method for the mass production of CNTs [73].

16.5.6.4.2 NANOTUBES FOR MOLECULAR TRANSPORT AND SEPARATIONS

Nanotubular materials are important "building blocks" of nanotechnology, based upon their incorporation into nanoscale devices. In particular, the synthesis and applications of carbon nanotubes have been extensively studied for two decades [74, 75]. One application area has been the use of carbon nanotubes for molecular separations, owing to some of their unique properties. One such important property, extremely fast mass transport of molecules within carbon nanotubes associated with their low friction inner nanotube surfaces, has been demonstrated via computational and experimental studies [76, 77].

Furthermore, the behavior of adsorbate molecules in nano-confinement is fundamentally different than in the bulk phase, which could lead to the design of new sorbents [78]. Finally, their one-dimensional geometry could allow for alignment in desirable orientations [79] for given separation devices to optimize the mass transport. Despite possessing such attractive properties, several intrinsic limitations of carbon nanotubes inhibit their application in large scale separation processes: the high cost of carbon nanotube synthesis and membrane formation (by microfabrication processes), as well as their lack of surface functionality, which significantly limits their molecular selectivity.

Although outer-surface modification of carbon nanotubes has been developed for nearly two decades, interior modification via covalent chemistry is still challenging due to the low reactivity of the inner-surface. Specifically, forming covalent bonds at inner walls of carbon nanotubes requires a transformation from sp^2 to sp^3 hybridization. The formation of sp^3 carbon is energetically unfavorable for concave surfaces. Hence, the interior functionalization of carbon nanotubes remains a challenge.

16.5.6.4.3　CARBON NANOTUBES (CNTS) SPONGE

CNTs are nanoscale structures with excellent mechanical and electronic properties. Production of macroscopic, engineered structures based on assembled CNTs with controlled orientation and configuration is an important step towards practical applications. To this end, CNT-based macrostructures have been fabricated in various forms such as vertically aligned arrays via direct growth, long fibers or sheets yarned from CNT forests [80, 81], solids and wafers densified from aligned films [82], solution-processed aerogels [83] and thin films or bucky papers [84], in which the alignment, density, and porosity of the CNTs can be tailored. These structures demonstrated potential applications in a variety of areas such as supercapacitors [82], catalytic electrodes [85], flexible heaters [82], superhydrophobic coatings [86], biomimetic surfaces [87], artificial muscles [88], and integrated microelectromechanical systems. The engineered nanostructures offer a great opportunity to develop high-performance materials and explore applications in various fields.

Most importantly, the light-weight, high porosity, and large surface area of CNT-based materials make them a promising candidate for environmental applications such as sorption, filtration and separation [89, 90]. For example, CNTs were considered as superior sorbents for a wide range of organic chemicals and inorganic contaminants than conventional systems such as clay and activated carbon, with a number of advantages including stronger chemical-nanotube interactions, rapid equilibrium rates, high sorbent capacity and tailored surface chemistry [90]. Filters based on aligned CNT arrays were used to remove hydrocarbons or bacterial, and separate different-size proteins [91, 92]. Membranes made from parallel CNT channels with tunable pore size and density have shown high flux for gas and water, and selective molecular transport through pore functionalization [93]. Despite the attractive physical properties of CNTs and intensive efforts in this area, the progresses in environmental applications have been limited in the sorption or separation of diluted molecules and ions in very small amount or scale [94].

The synthesis of a sponge-like bulk material consist of self-assembled, interconnected CNT skeletons, with a density close to the lightest aerogels, a porosity of >99%, high structural flexibility and robustness, and wettability to organics in pristine form. The CNT sponges can be deformed into any shapes elastically and compressed to large-strains repeatedly in air or liquids without collapse. The sponges in densified state swell instantaneously upon contact with organic solvents. They absorb a wide range of solvents and oils with excellent selectivity, recyclability, and absorption capacities up to 180 times their own weight, two orders of magnitude higher than activated carbon. A small densified pellet floating on water surface can quickly remove a spreading oil film with an area up to 800 times that of the sponge, suggesting potential environmental applications such as water remediation and large-area spill cleanup. In comparison, the application of one of the lightest porous materials, silica aerogel, has been impeded by their structural fragility, environ-mental sensitivity and high production cost [95, 96].

Carbon nanotubes have a unique one-dimensional structure, large specific surface area, and are oleophilic and hydrophobic [64]. As filter materials, CNTs exhibit selective separation characteristics for different solutions [65, 66]. The CNTs directly synthesized on the surface of expanded vermiculite could significantly improve the oil sorption capacity by forming a hydrophobic surface [67, 97].

Here we used a highly porous CNT material (CNT sponges) as sorbents for oil absorption. As shown in Fig. 16.7a–c, a CNT sponge can be considered as a three-dimensional scaffold, which is highly porous and intrinsically hydrophobic, but with a strong affinity for most organic solvents [98].

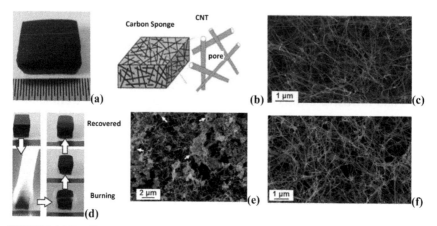

FIGURE 16.7 CNT sponges: (a) photograph, (b) schematic of the pore structure, (c) SEM image of the CNT sponge, (d) burning and reuse of the CNT sponge. SEM images of (e) the surface and (f) the core of the CNT sponge after burning.

CNT sponges have high sorption capacities and high sorption rates for many organics and oils, including mineral oil, vegetable oil and diesel oil. The sorption capacities are greater than 100 g g^{-1}. The sorption kinetics process has been described by a second-order kinetic model. After sorption, the CNT sponges are reusable upon burning or squeezing, and show good [99].

Figure 16.7a shows a photograph of the as-prepared CNT sponge. The CNT sponge is super-light, highly flexible and super-elastic. It possesses a unique porous structure as illustrated in Fig. 16.7b. SEM characterization shows that the CNT sponge consists of highly interconnected CNTs (Fig. 16.7c). The pore size ranges from several nanometers to tens of micrometers [98]. Owing to their high porosities, these CNT sponges can be densified into small balls by absorbing a minute amount of ethanol or acetone, making their storage and transportation simple and convenient. A balled-up CNT sponge will recover its initial shape and size after absorption of other organics or oil. After absorption, the absorbed oil or organics can be readily desorbed by directly burning the oil from within the sponge or squeezing the liquid out of the sponge. Figure 16.7d shows the burning process of a CNT sponge in air after diesel oil sorption. The oil contained in the sponge can be completely burned off from the CNT sponge without destroying the sponge structure due to the high thermal stability of CNTs.

Thermogravimetric analysis (TGA) indicates that the CNT sponge can sustain a temperature as high as 600 °C in air. After burning, the volume of the sponge reduced slightly, but quickly recovered to its initial volume when immersed into oil for reabsorption. SEM images show some burning residues left on the surface of the CNT sponge after burning (Fig. 16.7e). However, the core of the CNT sponge is very pure and clean, showing no obvious difference compared to the pristine sponge (Fig. 16.7f).[99]

The as-grown lowest-density (5.8 mg/cm³) sponge appears as a black carpet with an area of about 12 cm² and a thickness of 3.5 mm. This soft and flexible sponge can be manipulated into tight scrolls without splitting of structure (Fig. 16.8). Sponges with lower densities can be compacted more easily than larger density samples. When two sponge bricks with density of 5.8 and 11.6 mg/cm³, respectively, are placed in series to support a standard weight of 200 g, the smaller density brick shows much larger deformation under the same compressive force (Fig. 16.9) [100].

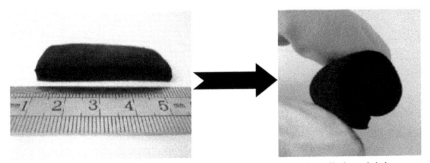

FIGURE 16.8 Pictures of a soft carpet like sponge, which can be rolled up tightly.

FIGURE 16.9 Pictures of two stacked sponges supporting a weight of 200 g, in which the top sponge was compressed to a larger degree due to its lower density (5.8 mg/cm³).

The CNT sponge can be synthesized through a CVD growth on a quartz substrate (Supplementary Section) [98]. Figure 16.10a–d. shows the macroscopic photo of the flexible and soft CNT sponge. The thickness of the CNT sponge is controlled by its growth time. The sponge can be easily peeled off from the quartz substrate after its growth, maintaining its pristine integral shape and structure (Fig. 16.10a). The self-sustainable structure and the robust mechanical properties of the CNT sponge are similar to other high-porosity materials (aerogels) or aligned CNT arrays [67, 101, 102]. In addition, CNT sponges are not only more flexible and soft, but they also show viscoelasticity and good elastic recovery. It can be bent into a large extent without collapse or breaking apart (Fig. 16.10b and 16.10c).

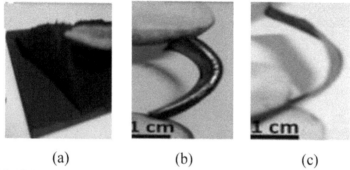

(a)   (b)   (c)

FIGURE 16.10   Macro-/micro-structures of CNT sponges: (a) the CNT sponges that peeled from the substrate; (b and c) Flexibility of CNT sponges with different thickness.

The CNT sponges display structural flexibility that is rarely observed in other high-porosity materials (aerogels) or aligned CNT arrays. Manual compression to >50% volume reduction shows that the sponge is compliant and springy (see Supporting Information Movie). A bulk sponge can be bent to a large degree or twisted without breaking apart, and after that still recover to nearly original shape. Prolonged ultra-sonication in solvents did not dismantle the sponges. The structural integrity under large deformations is owing to the highly interconnected CNTs in a three-dimensionally isotropic configuration, which could prevent the sliding or splitting between CNTs along any direction. In contrast, CNT aerogels prepared by solution processing and critical-point-drying are very brittle and need reinforcement by polymer infiltration [83].

By introducing solvents (ethanol) and then gently compressing the sponge to remove the solvent, we formed densified pellets in dry state with controlled shapes such as flat carpets or spherical particles (Fig. 16.11). There is about 10 to 20 times volume shrinkage during densification and much higher volume reduction is possible by processing in vacuum to make CNTs more densely packed. The densified pellets instantaneously swell to larger sizes by adding droplets of ethanol until recover to the original volume (Fig. 16.11) and can be densified again by squeezing out the absorbed solvent for many times, therefore the swelling process is fully reversible. Shape and structural recovery of the sponge stems from the random distribution of CNTs that prevents the formation of strong van der Waals interactions even at densified state, therefore liquid re-absorption into the pores could push CNTs away and back to their original configuration. For other structures such as aligned forests, introduction of liquids zip them into a densely packed solid due to maximum van der Waals forces produced between parallel CNTs, which is an irreversible process [82].

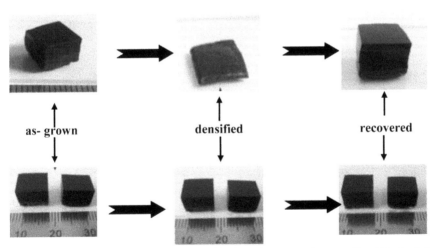

FIGURE 16.11 Densification of two cubic-shaped sponges into small pellets (a flat carpet and a spherical particle, respectively) and full recovery to original structure upon ethanol absorption.

The selective uptake of solvent (in preference to water) is important for environmental applications. An investigation onto sponges presented that the sponges show no strength degradation after compression at a set strain of 60% for 1000 cycles (Fig. 16.12), suggesting structural robustness compared with brittle silica aerogels. Interestingly, volume reduction of the sponges after they were compressed in solvent for 1000 times has not been observed (inset of Fig. 16.12), while a deformation of nearly 20% was measured during compression in air as shown in inset of Fig. 16.12. It shows that the absorption and filling of solvents into the pores is favorable for the sponges to recover the original structure. Reversible absorption and removal of solvents for many times also provides a simple way for recycled use.

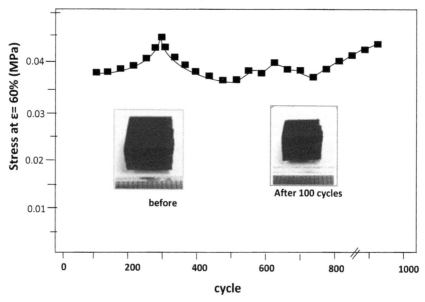

FIGURE 16.12 Stress recorded for 1000 cycles at set e¼60% showing no strength degradation. Inset, photos of the CNT sponge before and after cyclic tests.

Several studies of CNT sponges show that they demonstrate high absorption capacities (defined by Q, the ratio between the final and initial weight after full absorption) of 80 to 180 times their own weight for a wide range of solvents and oils, with larger Q for higher-density liquids (chloroform). The mechanism is mainly physical absorption of organic molecules, which can be stored in the sponge pores. Owing to the low sponge density and high porosity, the Q values are significantly higher than the weight gains of dense manganese oxide nanowire membranes (<20 times) made by filtration for similar solvents and oils [103].

The absorbed oil or solvents can be removed by mechanical compression or directly burned in air without destroying the sponge structure. Many porous materials with different pore sizes and densities including natural fibrous products (cotton towel, loofah), polymeric sponges (polyurethane- or polyester-based) and pellets of activated carbon have been investigated by several researchers. In case of diesel oil, the absorption capacity of CNT sponges (Q=143) is several times that of polymeric sponges (Q<40), 35 times that of cotton and loofah (Q<4) and two orders of magnitude higher than activated carbon (Q<1) (Fig. 16.13).

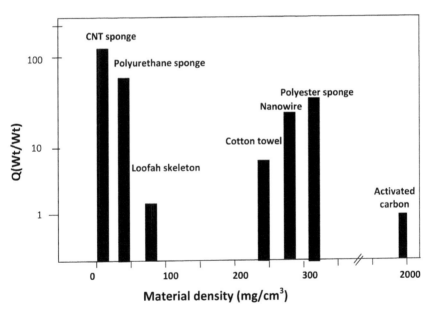

FIGURE 16.13 Summary of the absorption capacity for diesel oil measured from CNT sponges, natural products (cotton, loofah), polymeric sponges (polyurethane, polyester) and activated carbon. The maximum weight gain for oils by previously reported nanowire membranes (Q=20) [104] is included for comparison.

The CNT sponges in pristine or densified can actively absorb and remove different mount of oils spreading on water surface. It can be observed that the oil film kept shrinking toward center while the sponge grew to a larger size to oil absorption (Fig. 16.14). The growing size of the sponge accompanied by the shrinkage of oil film indicates continuous oil absorption and storage inside the sponge.

FIGURE 16.14 Snapshots showing the absorption of vegetable oil film (dyed with Oil Blue) distributed on a water bath by a small spherical sponge.

After uptake of the entire oil film, the sponge still floating on water has recovered to nearly original shape and size. In addition, a piece of pristine sponge can continuously attract and suck most part of an oil film when it was placed to contact the edge of the film (Fig. 16.15).

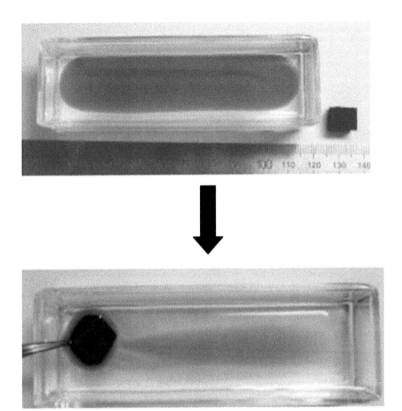

FIGURE 16.15   Active absorption of a continuous oil strip distributed in a rectangular water channel by a small non-densified sponge placed near the left edge and in contact with oil.

Significantly, a small particle of densified CNT sponge can remove a spreading diesel oil film. It can be observed that the sponge is floating on water surface and moving freely throughout the oil area. Wherever it arrives, the sponge instantaneously sucks the part of oil film in contact, resulting in a local white-color region around and behind where fresh water exposes. The sponge tends to drift to the remaining oil film area due to its water-repelling and oil-wetting properties, leading to this unique "floating-and-cleaning" capability that is particularly useful for spill cleanup. The sponge expanded after absorption of the whole oil film, leaving a clear water surface around. The oil area that has been completely cleaned is about 800 times larger than the size of the initial densified sponge (Fig. 16.16).

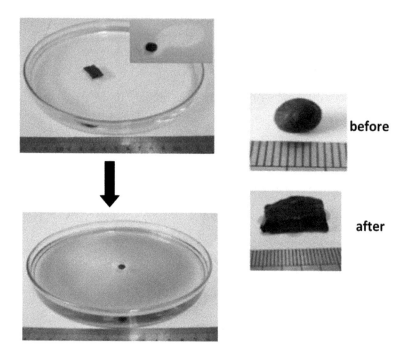

FIGURE 16.16   Large-area oil cleanup. (The oil area is about 800 times that of the projected sponge area. (middle) Clean water surface after complete oil absorption by the sponge, which has grown to a larger size and changed to a rectangular shape).

16.6   CONCLUDING REMARKS

This chapter introduced the main method establishing the oil spill model. We synthetically considered spreading and diffusion of oil slick, and taking the effect of evaporation, emulsification and sea adsorption into account. By considering different oil properties, the oil spill model was reviewed. Among various methods for removal oil, adsorption is one of the most effective ones. Many absorbent can be applied for this purpose. But unusual and special sorbent which has very considerable efficiency, is CNT sponges. So we discussed about CNT properties and the CNT sponges by using evidence mentioning.

KEYWORDS

- **Carbon nanotube sponges**
- **Oil properties**
- **Oil slick**
- **Oil spill**

REFERENCES

1. Al-Majed, A. A., Adebayo, A. R., & Hossain, M. E. (2012). *A sustainable approach to controlling oil spills.* Journal of Environmental Management, *113*, 213–227.
2. Diya'uddeen, B. H., et al. (2008). *Production of Activated Carbon from Corncobs and its Utilization in Crude Oil Spillage Clean Up* Agricultural Engineering International: the CIGR E-journal, 1–91.
3. Wang, J., Zheng, Y., & Wang, A. (2012). *Effect of Kapok Fiber Treated with Various Solvents on Oil Absorbency.* Industrial Crops and Products, *40*, 178–184.
4. Lim, T.-T., & Huang, X. (2007). *Evaluation of kapok (Ceiba pentandra (L.) Gaertn.) as a natural hollow hydrophobic–oleophilic fibrous sorbent for oil spill cleanup.* Chemosphere, *66(5)*, 955–963.
5. Lehr, W. J. (2001). *Review of Modeling Procedures for Oil Spill Weathering Behavior.* Advances in Ecological Sciences, *9*, 51–90.
6. Mackay, D., Stiver, W., & Tebeau, P. A. (1983). *Testing of Crude Oils and Petroleum Products for Environmental Purposes.* In *International Oil Spill Conference*: American Petroleum Institute.
7. Elliott, A. J., Hurford, N., & Penn, C. J. (1986). *Shear Diffusion and the Spreading of Oil Slicks.* Marine Pollution Bulletin, *17(7)*, 308–313.
8. Chen, H. Z., Lida, M., & Li, X. (2007). *Mathematical Modeling of Oil Spill on the Seaand Application of the Modeling in Daya Bay.* Hydrodynamics, *19(3)*, 282–291.
9. Zhao, W., Huang, Q & Luo, L. (1995). *Experimental Study on Diffusion of Oiled Wastewater under Water Surface.* Journal of Hydrodynamics, *10(6)*, 679–684.
10. Topham, D. R. (2002). *An Analysis of the Performance of Weir Type Oil Skimmers.* Spill Science & Technology Bulletin, *7(5–6)*, 289–297.
11. Durkee, J. (2008). *The Second Cleaning Process: The Oil Skimmer—Part II.* Metal Finishing, *106(7–8)*, 69–71.
12. Rudnitskii, E. M., & Selyakh, G. S. (1964). *A high-vacuum unit for oil-free operation with a magnetic discharge pump.* Vacuum, *14(10)*, 1144–1146.
13. Ghannam, M. T., & Chaalal, O. (2003). *Oil Spill Cleanup using Vacuum Technique.* Fuel, *82(7)*, 789–797.
14. Bitting, K. R. (1999). *Evaluating a Protocol for Testing a Fire-Resistant Oil-Spill Containment Boom.* Spill Science & Technology Bulletin, *5(5–6)*, 337–339.

15. Wong, K. V., & Barin, E. (2003). *Oil Spill Containment by a Flexible Boom System.* Spill Science & Technology Bulletin, *8(5–6),* 509–520.

16. Walz, L. A. (1999). *Second Phase Evaluation of a Protocol for Testing a Fire Resistant Oil Spill Containment Boom.* Spill Science & Technology Bulletin, *5(5–6),* 341–343.

17. Chapman, H., et al. (2007). *The Use of Chemical Dispersants to Combat Oil Spills at Sea: A Review of Practice and Research Needs in Europe.* Marine Pollution Bulletin, *54(7),* 827–838.

18. Karakasi, O. K., & Moutsatsou, A. (2010). *Surface Modification of High Calcium Fly Ash For Its Application in Oil Spill Clean Up.* Fuel, *89(12),* 3966–3970.

19. Banerjee, S. SJoshi, M. V., & Jayaram, R. V. (2006). *Treatment of Oil Spills Using Organo-Fly Ash.* Desalination, *195(1–3),* 32–39.

20. Yang, Y., et al. (2009). *Silkworms Culture as a Source of Protein for Humans in Space.* Advances in Space Research, *43(8),* 1236–1242.

21. Moriwaki, H., et al. (2009). Utilization of silkworm cocoon waste as a sorbent for the removal of oil from water. *Journal of Hazardous Materials, 165(1–3),* 266–270.

22. Lin, J., et al. (2012). *Nanoporous polystyrene fibers for oil spill cleanup.* Marine Pollution Bulletin, *64,* 347–352.

23. Zheng, Y. P., et al. (2004). *Sorption Capacity of Exfoliated Graphite for Oils-Sorption in and Among Worm-Like Particles.* Carbon, *42(12–13),* 2603–2607.

24. Merlin, F. X., & P. L. (2009). Guerroué, *Use of Sorbents for Spill Response.*

25. Annunciado, T. R., Sydenstricker, T. H. D. & Amico, S. C. (2005). *Experimental Investigation of Various Vegetable Fibers as Sorbent Materials for Oil Spills.* Marine Pollution Bulletin, *50(11),* 1340–1346.

26. Adebajo, M. O., et al. (2003). *Porous Materials for Oil Spill Cleanup: A Review of Synthesis and Absorbing Properties.* Journal of Porous Materials, *10(3),* 159–170.

27. Teas, C., et al. (2001). *Investigation of the Effectiveness of Absorbent Materials in Oil Spills Clean up.* Desalination, *140(3),* 259–264.

28. Sidik, S. M., et al. (2012). *Modified Oil Palm Leaves Adsorbent with Enhanced Hydrophobicity for Crude Oil Removal.* Chemical Engineering Journal, *203,* 9–18.

29. Fritt-Rasmussen, J., & Brandvik, P. J. (2011). *Measuring Ignitability for in Situ Burning of Oil Spills Weathered under Arctic Conditions: From Laboratory Studies to Large-Scale Field Experiments.* Marine Pollution Bulletin, *62(8),* 1780–1785.

30. Abdul Aziz Al-Majed., Abdulrauf, M. E. H., & Adebayo, R. (2012). *A Novel Sustainable Oil Spill Control Technology.* Environmental Engineering and Management Journal

31. Fingas, M. (2011). *Chapter 13 – Weather Effects on Oil Spill Countermeasures,* in *Oil Spill Science and Technology,* F. Mervin, Editor Gulf Professional Publishing: Boston, 339–426.

32. Barnea, N., *Health and Safety Aspects of In-situ Burning of Oil.* National Oceanic and Atmospheric Administration.

33. Buist, I., et al. (1999). In Situ Burning. *Pure and Applied Chemistry, 71(1),* 43–65.

34. Oebius, H. U. (1999). Physical Properties and Processes that Influence the Clean Up of Oil Spills in the Marine Environment. *Spill Science & Technology Bulletin, 5(3–4),* 177–289.

35. Jensen, H. V., & Mullin, J. V. (2003). MORICE—New Technology for Mechanical oil Recovery in Ice Infested Waters. *Marine Pollution Bulletin, 47(9–12),* 453–469.

36. Broje, V., & Keller, A. A. (2007). Effect of Operational Parameters on the Recovery Rate of an Oleophilic Drum Skimmer. *Journal of Hazardous Materials, 148(1–2),* 136–143.
37. Graham, P. (2010). Deep Sea Oil Spill Cleanup Techniques: Applicability, Trade-offs and Advantages. *Proquest Discovery Guides.*
38. Shashwat S. Banerjeea, Milind V. Joshi, & Jayarama, R. V. (2006). Treatment of oil spills using organo-fly ash. *Desalination,* 195. 32–39.
39. Carmody, O., et al. (2007). *Adsorption of hydrocarbons on organo-clays—Implications for oil spill remediation.* Journal of Colloid and Interface Science, *305(1),* 17–24.
40. Sugumaran, P., et al. (2012). *Production and Characterization of Activated Carbon from Banana Empty Fruit Bunch and Delonix regia Fruit Pod* Journal of Sustainable Energy and Environment, 3p. 125–132.
41. Alaya, M. N. (2000). *Activated Carbon from Some Agricultural Wastes Under Action of One-Step Steam Pyrolysis.* Journal of Porous Materials, 7. 509–517.
42. Srinivasakannan, C., & Zailani Abu Bakar, M. (2004). *Production of activated carbon from rubber wood sawdust.* Biomass and Bioenergy, *27(1),* 89–96.
43. Rambabu, N., et al. (2013). *Evaluation and Comparison of Enrichment Efficiency of Physical/Chemical Activations and Functionalized Activated Carbons Derived from Fluid Petroleum Coke for Environmental Applications.* Fuel Processing Technology, *106(0),* 501–510.
44. Kawano, T., et al. (2008). *Preparation of Activated Carbon from Petroleum Coke by KOH Chemical Activation for Adsorption Heat Pump.* Applied Thermal Engineering, *28(8–9),* 865–871.
45. Li, X., et al. (2009). *Catalytic Ozonation of P-Chlorobenzoic Acid by Activated Carbon and Nickel Supported Activated Carbon Prepared from Petroleum Coke.* Journal of Hazardous Materials, *163(1),* 115–120.
46. Gupta, V. K., et al. (2011). *Pesticides Removal from Waste Water by Activated Carbon Prepared from Waste Rubber Tire.* Water Research, *45(13),* 4047–4055.
47. San Miguel, G., Fowler, G. D., & Sollars, C. J. (2002). *The leaching of inorganic species from activated carbons produced from waste tire rubber.* Water Research, *36(8),* 1939–1946.
48. San Miguel, G., Fowler, G. D., & Sollars, C. J. (2003). *A Study of the Characteristics of Activated Carbons Produced by Steam and Carbon Dioxide Activation of Waste Tyre Rubber.* Carbon, *41(5),* 1009–1016.
49. Hussein, M., et al. (2011). *Heavy Oil Spill Cleanup using Law Grade Raw Cotton Fibers: Trial for Practical Application* Petroleum Technology and Alternative Fuels, *2(8),* 132–140.
50. Bayat, A., et al. (2005). *Oil Spill Cleanup from Sea Water by Sorbent Materials.* Chemical Engineering & Technology, *28(12),* 1525–1528.
51. Sun, X. F., Sun, R. C., & Sun, J. X. (2003). *A Convenient Acetylation of Sugarcane Bagasse using NBS as a Catalyst for the Preparation of Oil Sorption-Active Materials.* Journal of materials science, 38, 3915–3923.
52. Husseien, M., et al. (2009). *A Comprehensive Characterization of Corn Stalk and Study of Carbonized Corn Stalk in Dye and Gas Oil Sorption.* Journal of Analytical and Applied Pyrolysis, *86(2),* 360–363.

53. Hussein, M., Amer, A. A., & Sawsan, I. I. (2008). *Oil Spill Sorption using Carbonized Pith Bagasse: 1. Preparation and Characterization of Carbonized Pith Bagasse.* Journal of Analytical and Applied Pyrolysis, *82(2),* 205–211.

54. Inagaki, M., A. Kawahara., & Konno, H. (2002). *Sorption and Recovery of Heavy Oils using Carbonized Fir Fibers and Recycling.* Carbon, *40(1),* 105–111.

55. Chol, H. M. (1992). *Natural Sorbents in Oil Spill Cleanup* Environmental Science and Technology, *26,* 772–776.

56. Suni, S., et al. (2004). *Use of a By-Product of Peat Excavation, Cotton Grass Fiber, as a Sorbent for Oil-Spills.* Marine Pollution Bulletin, *49(11–12),* 916–921.

57. Radetic, M., et al. (2008). *Efficiency of Recycled Wool-Based Nonwoven Material for the Removal of Oils from Water.* Chemosphere, *70(3),* 525–530.

58. Rajakovic, V., et al. (2007). *Efficiency of Oil Removal from Real Wastewater with Different Sorbent Materials.* Journal of Hazardous Materials, *143(1–2),* 494–499.

59. Srinivasan, A., & Viraraghavan, T. (2010). *Oil Removal from Water using Biomaterials.* Bioresource Technology, *101(17),* 6594–6600.

60. Ahmad, A. L., Sumathi, S., & Hameed, B. H. (2005). *Residual Oil and Suspended Solid Removal using Natural Adsorbents Chitosan, Bentonite and Activated Carbon: A Comparative Study.* Chemical Engineering Journal, *108(1–2),* 179–185.

61. Likon, M., et al. (2012). *Populus seed fibers as a natural source for production of oil super absorbents.* Journal of Environmental Management,

62. Wang, J., Zheng, Y., & Wang, A. (2012). *Superhydrophobic Kapok Fiber Oil-Absorbent: Preparation and High Oil Absorbency.* Chemical Engineering Journal, *213,* 1–7.

63. Abdullah, M. A., Rahmah, A., & Man, Z. (2010). *Physicochemical and Sorption Characteristics of Malaysian Ceiba pentandra (L.) Gaertn. as a Natural Oil Sorbent.* Journal of Hazardous Materials, *177,* 683–691.

64. Zhang, Y., et al. (2009). *A Double-Layered Carbon Nanotube Array with Super-Hydrophobicity.* Carbon, *47(14),* 3332–3336.

65. Sears, K., et al. (2010). *Recent Developments in Carbon Nanotube Membranes for Water Purification and Gas Separation* Materials, *3,* 127–149.

66. Wu, Q., et al. (2010). *Study of Fire Retardant Behavior of Carbon Nanotube Membranes and Carbon Nanofiber Paper in Carbon Fiber Reinforced Epoxy Composites.* Carbon, *48(6),* 1799–1806.

67. Hashim, D. P., et al. (2012). *Covalently Bonded Three-Dimensional carbon Nanotube Solids via Boron Induced Nanojunctions.* Carbon nanotubes and fullerenes,

68. Kroto, H. W., et al. (1985). *C 60: Buckminsterfullerene.* Nature, *318(6042),* 162–163.

69. Iijima, S. (1991). *Helical Microtubules of Graphitic Carbon.* nature, *354(6348),* 56–58.

70. Novoselov, K. S., et al. (2004). *Electric Field Effect in Atomically thin Carbon Films.* Science, *306(5696),* 666–669.

71. Geim, A. K., & Novoselov, K. S. (2007). *The Rise of Graphene.* Nature materials, *6(3),* 183–191.

72. Dresselhaus, M. S., Dresselhaus, G., & Eklund, P. C. (1996). *Science of Fullerenes and Carbon Nanotubes: Their Properties and Applications*, Academic Press. 965.

73. Huang, J., et al. (2012). *A Review of the Large-Scale Production of Carbon Nanotubes: The Practice of Nanoscale Process Engineering.* Chinese Science Bulletin, *57(2–3),* 157–166.

74. Lau, K., Gu, C., & Hui, D. (2006). *A Critical Review on Nanotube and Nanotube/ Nanoclay Related Polymer Composite Materials.* Composites Part B: Engineering, *37(6)*, 425–436.

75. Choi, W., et al. (2010). *Carbon Nanotube-Guided Thermopower Waves.* Materials To-day, *13(10)*, 22–33.

76. Sholl, D. S., & Johnson, J. K. (2006). *Making High-Flux Membranes with Carbon Nanotubes.* Science, *312(5776)*, 1003–1004.

77. Zang, J., et al. (2009). *Self-Diffusion of Water and Simple Alcohols in Single-Walled Aluminosilicate Nanotubes.* ACS nano, *3(6)*, 1548–1556.

78. Talapatra, S., Krungleviciute, V., & Migone, A. D. (2002). *Higher Coverage Gas Ad-sorption on the Surface of Carbon Nanotubes: Evidence for a Possible new Phase in the Second Layer.* Physical review letters, *89(24)*, 246106.

79. Mauter, M. S., Elimelech, M., & Osuji, C. O. (2010). *Nanocomposites of Vertically Aligned Single-Walled Carbon Nanotubes by Magnetic Alignment and Polymerization of a Lyotropic Precursor.* ACS nano, *4(11)*, 6651–6658.

80. Zhang, M., Atkinson, K. R., & Baughman, R. H. (2004). *Multifunctional Carbon Nanotube Yarns by Downsizing an Ancient Technology.* Science, *306(5700)*, 1358–1361.

81. Zhang, M., et al. (2005). *Strong, Transparent, Multifunctional, Carbon Nanotube Sheets.* Science, *309(5738)*, 1215–1219.

82. Futaba, D. N., et al. (2006). *Shape-Engineerable and Highly Densely Packed Single-Walled Carbon Nanotubes and Their Application as Super-Capacitor Electrodes.* Na-ture materials, *5(12)*, 987–994.

83. Bryning, M. B., et al. (2007). *Carbon Nanotube Aerogels.* Advanced Materials, *19(5)*, 661–664.

84. Campos-Delgado, J., et al. (2008). *Bulk Production of a New form of sp2 Carbon: Crystalline Graphene Nanoribbons.* Nano letters, *8(9)*, 2773–2778.

85. Trancik, J. E., Barton, S. C., & Hone, J. (2008). *Transparent and Catalytic Carbon Nanotube Films.* Nano letters, *8(4)*, 982–987.

86. Lau, K., et al. (2003). *Superhydrophobic Carbon Nanotube Forests.* Nano letters, *3(12)*, 1701–1705.

87. Qu, L., et al. (2008). *Carbon Nanotube Arrays with Strong Shear Binding-on and easy Normal Lifting-Off.* Science, *322(5899)*, 238–242.

88. Aliev, A. E., et al. (2009). *Giant-Stroke, Superelastic Carbon Nanotube Aerogel Mus-cles.* Science, *323(5921)*, 1575–1578.

89. Mauter, M. S., & Elimelech, M. (2008). *Environmental Applications of Carbon-Based Nanomaterials.* Environmental Science & Technology, *42(16)*, 5843–5859.

90. Pan, B., & Xing, B. (2008). *Adsorption Mechanisms of Organic Chemicals on Carbon Nanotubes.* Environmental Science & Technology, *42(24)*, 9005–9013.

91. Li, X., et al. (2007). *Compression-Modulated Tunable-Pore Carbon-Nanotube Mem-brane Filters.* Small, *3(4)*, 595–599.

92. Srivastava, A., et al. (2004). *Carbon Nanotube Filters.* Nature materials, *3(9)*, 610–614.

93. Yu, M., et al. (2008). *High Density, Vertically-Aligned Carbon Nanotube Membranes.* Nano letters, *9(1)*, 225–229.

94. Chen, W., Duan, L., & Zhu, D. (2007). *Adsorption of Polar and Nonpolar Organic Chemicals to Carbon Nanotubes.* Environmental Science & Technology, *41(24)*, 8295–8300.

95. Leventis, N., et al. (2002). *Nanoengineering Strong Silica Aerogels.* Nano letters, *2(9)*, 957–960.

96. Capadona, L. A., et al. (2006). *Flexible, Low-Density Polymer Crosslinked Silica Aerogels.* Polymer, *47(16)*, 5754–5761.

97. Huang, J. Q., et al. (2012). *A Review of the Large-Scale Production of Carbon Nanotubes: The Practice of Nanoscale Process Engineering.* Chinese Science Bulletin, *57(2–3)*, 157–166.

98. Gui, X., et al. (2010). *Soft, Highly Conductive Nanotube Sponges and Composites with Controlled Compressibility*, 4, 2320–2326.

99. Gui, X., et al. (2011). *Recyclable Carbon Nanotube Sponges for Oil Absorption.* Acta Materialia, *59(12)*, 4798–4804.

100. Gui, X., et al. (2010). *Soft, Highly Conductive Nanotube Sponges and Composites with Controlled Compressibility.* ACS nano, *4(4)*, 2320–2326.

101. Kimizuka, O., et al. (2008). *Electrochemical Doping of Pure Single-Walled Carbon Nanotubes used as Supercapacitor Electrodes.* Carbon, *46(14)*, 1999–2001.

102. Pushparaj, V., et al. (2012). *Deformation and Capillary Self-Repair of Carbon Nanotube Brushes.* Carbon, *50(15)*, 5618–5620.

103. Yuan, J., et al. (2008). *Superwetting Nanowire Membranes for Selective Absorption.* Nature Nanotechnology, *3(6)*, 332–336.

104. Cao, A., et al. (2005). *Super-Compressible Foam like Carbon Nanotube Films.* Science, *310(5752)*, 1307–1310.

CHAPTER 17

THERMAL STABILITY OF ELASTIC POLYURETHANE

I. A. NOVAKOV, M. A. VANIEV, D. V. MEDVEDEV,
N. V. SIDORENKO, G. V. MEDVEDEV, and D. O. GUSEV

CONTENTS

ABSTRACT

Thermal stability of polyurethane elastomers based on a product of the anionic copolymerization of butadiene and isoprene in the ratio of 80:20 and isoprene was first studied by DSC. The preferred conditions (temperature of the isothermal segment and oxygen consumption) were revealed to determine the oxidation induction time of this type of materials. The effect of Irganox 1010, Evernox 10, Songnox 1010 and 1010 Chinox stabilizers on the oxidation induction time has been studied.

It was found that the products of the same chemical structure, depending on commercial brands, may display different antioxidant activity in OIT tests. It was pointed out that this factor must be taken into account in developing polyurethane composition formulations.

17.1 INTRODUCTION

Polyurethane elastomers (PUE) are of great practical importance in various fields [1]. In particular, in developing PUE of molding compositions for sports and roofing the liquid rubbers (oligomers) of diene nature with a molecular weight of 2000–4000 are widely used as a polyol component. Usually these are homopolymers of butadiene and isoprene, the products of copolymerization of butadiene with isoprene or butadiene with piperylene and isocyanate prepolymers based on these oligomers.

After curing, the materials exhibit good physical-mechanical, dynamic and relaxation properties, high hydrolytic stability [2, 3]. However, the disadvantage of these PUEs is their low resistance to thermal-oxidation aging, due to the presence of double bonds in the oligomer molecules. Under the effect of weather conditions, irreversible changes leading to partial or complete loss of the fundamental properties and materials reduced lifetime take place.

To minimize these negative effects the stabilizers and antioxidants are most commonly used. However, traditional methods of evaluating the effectiveness of a stabilizer within the PUE require lengthy field tests or the materials exposure to high air temperatures for a period of scores of hours and several days [4].

Modern methods of thermal analysis can significantly reduce the time of polymer tests, and informative results, their accuracy and capability to forecast the coating lifetime are significantly improved and expanded [5–7]. In particular, the determination of oxidation induction time (OIT) and the oxidation onset temperature (OOT) by differential scanning calorimetry (DSC) is effective for the accelerated study of thermal-oxidation polymers stability. This rapid method has been recommended [8–10] and used for polyolefins [11–15], oils and hydrocarbons [16, 17] and PVC [18].

Information on the use of OIT method for polyurethanes is currently limited. There are only some patent data [19] and publications on the results of determination of OIT and OOT for automotive coating materials, derived from polyurethanes of simple and complex polyester structure [20]. There are actually no publications on the test techniques and results of evaluating the thermal oxidation stability of PUE based on diene oligomers by using DSC. In addition, it should be noted that there are quite a number of manufacturers of commercial stabilizers in the market. Experience has proven that even with the same chemical structure their efficacy may vary. For this reason, when formulating the composition and selecting the stabilizer, or when making a decision on the feasibility and acceptability of direct replacement of one brand by another, one must use a modern rapid method that would quickly assess and predict the thermal stability of the coating material.

In view of the above, the purpose of the present research is to study, by using DSC, the oxidation stability of PUE samples derived from butadiene and isoprene copolymer, and comparative assessment of OIT performance in the presence of different brands of pentaerythritol tetrakis[3-(3',5'-di-tert-butyl-4'-hydroxyphenyl)propionate] stabilizer.

17.2 EXPERIMENTAL PART

To obtain PUE we used an oligomer, which is a product of the anionic copolymerization of butadiene and isoprene in the ratio of 80:20. The molecular weight of 3200. Mass fraction of hydroxyl groups was 1%, and the oligomer functionality on them was 1.8.

The compositions were being prepared in a ball mill from 12 to 15 h. Homogenization of the components was carried out to the degree of grinding equal to 65. All formulations contained the same amount of the following ingredients: the above oligomer, filler (calcium carbonate), plasticizer of a complex ester nature, desiccant (calcium oxide), organic red pigment (FGR CI 112, produced by Ter Hell & Co Gmbh.).

Stabilizing agents varied in the amounts of 0.2, 0.6, and 1.0 wt. parts to 100 oligomer weight parts. The compositions were numbered in accordance with Table 17.1. Comparison sample was the material under the code 0, which did not contain the stabilizer.

TABLE 17.1   Brands and Content of the Stabilizers Used

PUE sample number	Stabilizer brand and content (per oligomer 100 weight parts)			
	Irganox 1010	Evernox 10	Songnox 1010	Chinox 1010
0	–	–	–	–
1	0.2	–	–	–
2	0.6	–	–	–
3	1	–	–	–
4	–	0.2	–	–
5	–	0.6	–	–
6	–	1	–	–
7	–	–	0.2	–
8	–	–	0.6	–
9	–	–	1	–
10	–	–	–	0.2
11	–	–	–	0.6
12	–	–	–	1

The compositions were cured taking into account the general content of the hydroxyl groups in the system under the action of the estimated volume of Desmodur 44 V20L polyisocyanate (Bayer Material Science AG). Mass fraction of isocyanate groups in the product was 32%. Chain branching agent was chemically pure glycerine, and the catalyst was dibutyl tin dilaurate (manufactured by "ACIMA Chemical Industries Limited Inc.").

Curing conditions: standard laboratory temperature and humidity, duration – 72 h.

Pentaerythritol tetrakis[3-(3',5'-di-tert-butyl-4'-hydroxyphenyl)pro-pionate] of Irganox 1010, Evernox 10, Songnox Chinox 1010 and 1010 trademarks was used as a stabilizer. Structural formula is shown in Fig. 17.1.

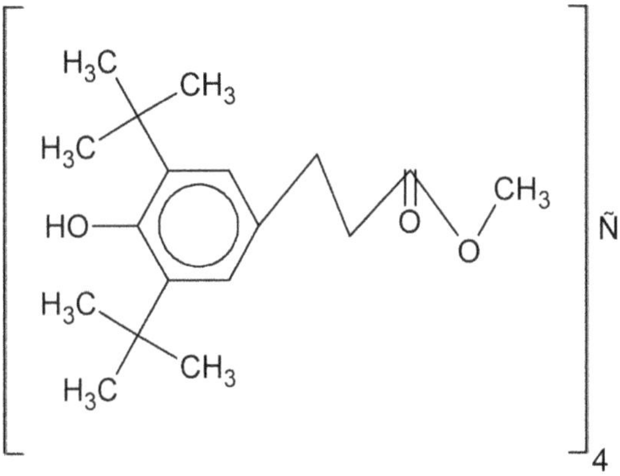

FIGURE 17.1   The structural formula of pentaerythritol tetrakis[3-(3',5'-di-tert-butyl-4'-hydroxyphenyl)propionate] stabilizer.

Samples of cured PUE were tested by the Nietzsche DSC 204 F1 Phoe-nix heat flow differential scanning calorimeter. Calibration was performed on an indium standard sample. Samples weighing 9 to 12 mg were placed in an open aluminum crucible. Test temperatures were attained at 10K/min rate under constant purging with an inert gas (argon). Upon reaching the target temperature the inert gas supply was stopped and the oxygen sup-ply started at 50 mL/min rate. All data were recorded and processed using Nietzsche Proteus special software in OIT registration mode.

17.3   RESULTS AND DISCUSSION

For polyolefins the tests to determine OIT are standardized in terms of both the recommended oxidizing gas (oxygen) flow and the isothermal segment temperature [8, 9]. There are no such standards for PUE. The authors [21] recommend that the test temperature be previously identi-fied experimentally, and other settings be selected in accordance with the recommendations of the relevant ASTM or ISO. For this reason, we first determined the conditions of OIT fixation for the non-stabilized sample at two different temperatures (Fig. 17.2).

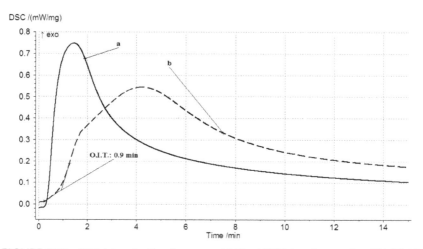

FIGURE 17.2   OIT determination for a non-stabilized PUE (number zero) at 200 °C (a) and 180 °C (b) by using atmosphere oxygen

As Fig. 17.2 shows, isothermal mode at 200 °C does not allow estimating the value of OIT for unstabilized sample. Under these conditions, snowballing oxidation degradation (curve a) starts almost immediately and oxidation induction time cannot be actually defined. As a result of reducing temperature to 180 °C it is possible to fix the target parameter. For unstabilized sample the OIT value was 0.9 min (curve b). However, we found that tests at lower temperatures lead to the significant increase in test duration, especially for stabilized samples. This is undesirable, since the benefit of rapid OIT method in this case is partially lost. Thus, we found the necessary balance between time and temperature conditions that allow estimating OIT for the investigated objects at the recommended oxygen supply rate. In this regard, all subsequent tests were carried out at 180 °C, and the results obtained are illustrated in each case, depending on the stabilizer type and content in comparison with the non-stabilized sample. The sample numbering and the stabilizer amount are consistent with Table 17.1.

Figure 17.3 shows the DSC curves for the samples containing Irganox 1010.

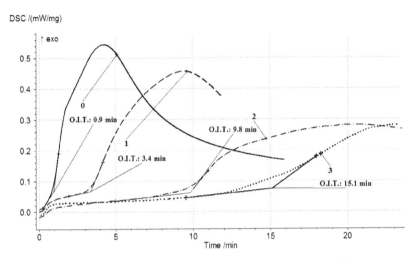

FIGURE 17.3   Isothermal DSC scans at 180°C for samples containing different amounts of Irganox 1010 stabilizer.

On DSC-curves of the materials stabilized by Irganox 1010 antioxidant, the OIT value changes can be traced depending on the content of pentaerythritol   tetrakis[3-(3',5'-di-tert-butyl-4'-hydroxyphenyl)propionate] of this brand. Significant stabilizing effect can be observed even at proportion of 0.2 weight parts. When adding 0.6 and 1.0 weight parts of this product to PUE the OIT was 9.8 and 15.1 min, respectively, which is 10 and 15 times as large as the corresponding parameter of reference sample.

Thereby it should be noted that the detected effects confirm and significantly specify the data obtained earlier [22] by using oligodiendiols and substituted phenol of this particular type, but using the classical evaluation method. The principal difference is that in the latter case, the implementation of the standard involves the need of thermostatic control of samples at higher air temperature (usually within 72 h), the subsequent physical and mechanical testing and correlation of properties before and after thermal aging. To assess the PUE sample oxidation stability by evaluating OIT by means of DSC the time expenditure is no more than 30 min and requires very little sample weight.

Materials containing Evernox 10, Songnox Chinox 1010 and 1010 were also investigated by this method. DSC data are shown in Figs. 17.4–17.6.

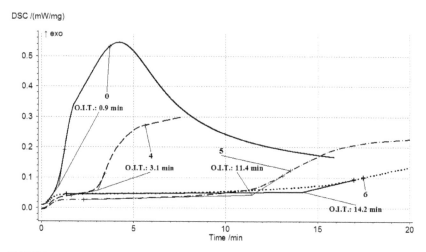

FIGURE 17.4 Isothermal DSC scans at 180°C for samples containing different amounts of Evernox 10 stabilizer.

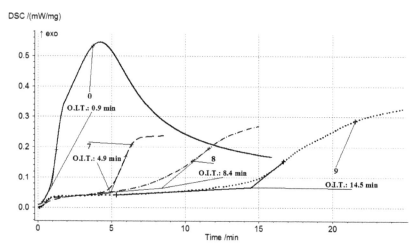

FIGURE 17.5 Isothermal DSC scans at 180°C for samples containing different amounts of Songnox 1010 stabilizer.

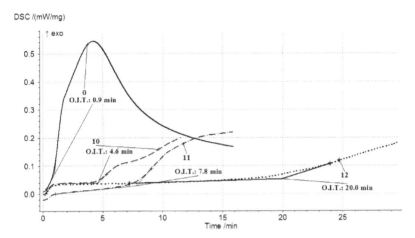

FIGURE 17.6 Isothermal DSC scans at 180°C for samples containing different amounts of Chinox 1010 stabilizer.

For general benchmarking the data obtained by processing the experimental array, are summarized in Table 17.2.

TABLE 17.2 OIT Values for PUE, Depending on the Type and Content of the Stabilizer

Sample number	Stabilizer brand and content (per 100 weight parts)				OIT, min
	Irganox 1010	Evernox 10	Songnox 1010	Chinox 1010	
0	–	–	–	–	0.9
1	0.2	–	–	–	3.4
2	0.6	–	–	–	9.8
3	1	–	–	–	15.1
4	–	0.2	–	–	3.1
5	–	0.6	–	–	11.4
6	–	1	–	–	14.2
7	–	–	0.2	–	4.9
8	–	–	0.6	–	8.4
9	–	–	1	–	14.5
10	–	–	–	0.2	4.6
11	–	–	–	0.6	7.8
12	–	–	–	1	20

It follows from the OIT numerical values that regardless of the stabilizer brand, with an increase of its content in PUE within the investigated concentration range, the natural increase in the oxidation induction time is recorded. However, due to the high sensitivity of the method a significant difference in stabilizing effect of equal quantities of the single-type product manufactured by different vendors can be easily traced. Simple calculation shows that the deviation between the OIT maximum and minimum values for materials stabilized by 0.2 weight parts of Irganox 1010, Evernox 10, Songnox Chinox 1010 and 1010, is equal to 36.7%. With the content being 0.6 and 1.0 weight parts, this deviation is 20.4 and 31.5%, respectively. Apparently, this difference is due to the chemical purity and other factors that determine the protective capacity of the products used.

17.4 CONCLUSIONS

Thus, in the example of mesh polyurethane materials based on copolymer of butadiene and isoprene we show the high efficiency of using the option of OIT determination by DSC in order to carry out express tests on PUE thermal stability. The preferred temperature of the isothermal segment for accelerated test of this type of materials has been deduced from experiment.

Comparative evaluation of OIT indicators for PUE, stabilized by pentaerythritol tetrakis[3-(3',5'-di-tert-butyl-4'-hydroxyphenyl)propionate], depending on the manufacturer, ceteris paribus, revealed significant difference in terms of the protective effect of sterically hindered phenol. In practical terms, this means that prior to making the composition and selecting the stabilizer, as well as planning direct qualitative and quantitative replacement of one brand stabilizer by another in the PUE, one must take into consideration the potential significant differences in antioxidant efficacy of the products.

17.5 ACKNOWLEDGEMENT

This work was supported by the Grant Council of the President of the Russian Federation, grant MK-4559.2013.3.

KEYWORDS

- **DSC**
- **Material testing**
- **Oxidation induction time**
- **Polyurethane elastomers**
- **Stabilizers**

REFERENCES

1. Prisacariu, C. (2011). Polyurethane Elastomers. From Morphology to Mechanical Aspects., *Springer-Verlag, Wien.*
2. Novakov, I. A., Nistratov, A. V., Medvedev, V. P., Pyl'nov, D. V., Myachina, E. B., Lukasik, V. A., et al. (2011). Influence of hardener on physicochemical and dynamic properties of polyurethanes based on α, ω-di(2-hydroxypropyl)-polybutadiene Krasol LBH-3000. *Polymer Science – Series D, 4(2),* 78–84.
3. Novakov, I. A., Nistratov, A. V., Pyl'nov, D. V., Gugina, S. Y., & Titova, E. N. (2012). Investigation of the effect of catalysts on the foaming parameters of compositions and properties of elastic polydieneurethane foams, *Polymer Science – Series D, 5(2),* 92–95.
4. ISO 188 (2011). Rubber, vulcanized or thermoplastic. Accelerated aging and heat resistance tests.
5. Joseph, D., Menczel, R., Bruce Prime., & John Wiley & Sons. (2009). *Thermal analysis of polymers. Fundamentals and Applications.* Inc., Hoboken, New Jersey.
6. Paul Gabbott. (2008*).* Principles and Applications of Thermal Analysis. Edited by Blackwell *Pub, Oxford, Ames, Iowa.*
7. Pieichowski, J., & Pielichowski, K. (1995). Application of thermal analysis for the investigation of polymer degradation processes, *J. Therm. Anal, 43,* 505–508.
8. ASTM D 3895-07: *Standard Test Method for Oxidative-Induction Time of Polyolefins by Differential Scanning Calorimetry.*
9. ISO 11357-6: *Differential scanning calorimetry (DSC).* Determination of oxidation induction time (isothermal OIT) and oxidation induction temperature (dynamic OIT).
10. ASTM E2009-08: Standard Test Method for Oxidation Onset Temperature of Hydrocarbons by *Differential Scanning Calorimetry.*
11. Gomory, I., & Cech, K. (1971). A new method for measuring the induction period of the oxidation of polymers, *J. Therm. Anal, 3,* 57–62.
12. Schmid, M., Ritter, A., & Affolter, S. (2006). Determination of oxidation induction time and temperature by DSC, *J. Therm. Anal. Cal, 83–2,* 367–371.
13. Woo, L., Khare, A. R., Sandford, C. L., Ling, M. T. K., & Ding, S. Y. (2001). Relevance of high temperature oxidative stability testing to long-term polymer durability, *J. Therm. Anal. Cal, 64,* 539–548.

14. Peltzer, M., & Jimenez, A. (2009). Determination of oxidation parameters by DSC for polypropylene stabilized with hydroxytyrosol (3,4-dihydroxy-phenylethanol), *J. Therm. Anal. Cal, 96(I)*, 243–248.
15. Focke, Walter W., & Westhuizen Isbe van der. (2010). Oxidation induction time and oxidation onset temperature of polyethylene in air, *J. Therm. Anal. Cal, 99*, 285–293.
16. Simon, P., & Kolman, L. (2001). DSC study of oxidation induction periods, *J. Therm. Anal. Cal, 64*, 813–820.
17. Conceicao Marta, M., Dantas Manoel, B., Rosenhaim Raul, Fernandes, Jr., Valter, J., Santos Ieda, M. G., & Souza Antonio, G. (2009). Evaluation of the oxidative induction time of the ethylic castor biodiesel, *J. Therm. Anal. Cal, 97*, 643–646.
18. Woo, L., Ding, S. Y., Ling, M. T. K., & Westphal, S. P. (1997). Study on the oxidative induction test applied to medical polymers, *J. Therm. Anal, 49*, 131–138.
19. Dietmar Mäder (Oberursel, DE), inventors. (2008). Stabilization of polyol or polyurethane compositions against thermal oxidation, US20090137699, USA,
20. Simon, P., Fratricova, M., Schwarzer, P., & Wilde, H.-W. (2006). Evaluation of the Residual Stability of Polyuretane Automotive Coatings by DSC, *J. Therm. Anal. Cal, 84(3)*, 679–692.
21. Clauss, M., Andrews, S. M., & Botkin, J. H. (1997). Antioxidant Systems for Stabilization of Flexible Polyurethane Slabstock, *J. Cellular Plastics, 33*, 457.
22. Medvedev, V. P., Medvedev, D. V., Navrotskii, V. A., & Lukyanichev, V. V. (2007). The study of oxidative aging polydieneurethane, *Polyurethane Technologies, 3*, 34–36.

INDEX

N

Nanofiller particles, 54, 55, 57–59, 61, 63, 66, 67, 70–73, 80, 81, 88
Nanoindentation method, 79
Nanoparticle, 62, 63, 66, 67, 69–73, 82, 188
Nano-porous structure, 101
Nanoscale devices, 229
Nanosized titanium dioxide, 189
Nanosized η-modification, 188
Narrowed crystallization peak, 193
Natural dilution, 220
Natural disasters, 201
    earthquakes, 201
    hurricanes, 201
Natural environmental conditions, 188
    nontoxicity, 188
Natural fibrous products, 237
Natural organic sorbents, 224
Natural seeps, 198
N-heptane, 14, 17, 148
Nickel filter, 29
Nietzsche Proteus, 251
Nigerian petroleum industry, 201
Nitroxyl radicals, 4–6
N-nitrosodiphenylamine, 158
Non-concentrated latex, 160
Noncrystalline phase, 4
Noncrystalline, 7, 8, 10
Non-native microorganisms, 219
    bio-augmentation, 219
Non-Newtonian fluid, 208
Non-staining antioxidant, 160
    phenolic type, 160
Non-woven fabric, 191
Novichem, 95
Nucleating agent, 193
Nucleation, 45

O

Offshore oil explorations, 201
Offshore platforms, 198
Oil drilling operations, 201
Oil industry, 199, 214
Oil spill cleanup technology, 219
Oil-consuming bacteria, 201

acid-producing bacteria, 201
general aerobic bacteria, 201
sulfate-reducing bacteria, 201
Oiliness emulsion, 211, 212
Oil-laden material, 221
Oil-soaked birds survive, 201
Oil-spill countermeasures, 213
Oil-wetting properties, 239
OIT numerical values, 256
OIT registration mode, 251
Oliver-pharr method, 79
One-dimensional geometry, 229
Open-cell polyurethane foams, 223
Operational resistance, 110
Optical microscope, 55, 67, 79, 190
Optimal properties, 131
Organic peroxide, 131
Organic solvents, 159, 231
Organic substance, 188
Organophilic clays, 222
Original latexes, 162
Original papers, 189
Orthorhombic, 6
Oxidation induction time, 248, 249, 252, 256
    1010 Chinox, 248
    Evernox 10, 248
    Irganox 1010, 248
    Songnox 1010, 248
Oxidation onset temperature, 249
Oxygen consumption, 248
Oxygen plasma activation, 145
    kaolin, 145
    silica, 145
Ozone, 14, 122, 158–165
    electrosynthesis, 161
    resistance, 122
Ozone-air mixtures, 159
Ozonide cycles, 164
Ozonization degree, 158, 159, 162

P

Patent declarations, 144
Pentaerythritol tetrakis, 249, 251, 253, 256
Permanent elongation, 113
Peroxide radicals, 104, 131